# SAVE ENERGY SAVE MONEY

## Simple Steps to Slash Your Energy Use and Costs

by Alvin Ubell and George Merlis

ALPHA

A Pearson Education Company

Copyright © 2002 by Alvin Ubell and George Merlis

All rights reserved. No part of this book shall be reproduced, stored in a retrieval system, or transmitted by any means, electronic, mechanical, photocopying, recording, or otherwise, without written permission from the publisher. No patent liability is assumed with respect to the use of the information contained herein. Although every precaution has been taken in the preparation of this book, the publisher and authors assume no responsibility for errors or omissions. Neither is any liability assumed for damages resulting from the use of information contained herein. For information, address Alpha Books, 201 West 103rd Street, Indianapolis, IN 46290.

International Standard Book Number: 0-02-864279-1
Library of Congress Catalog Card Number: 2001095857

04  03  02     8  7  6  5  4  3  2  1

Interpretation of the printing code: The rightmost number of the first series of numbers is the year of the book's printing; the rightmost number of the second series of numbers is the number of the book's printing. For example, a printing code of 02-1 shows that the first printing occurred in 2002.

*Printed in the United States of America*

**Note:** This publication contains the opinions and ideas of its authors. It is intended to provide helpful and informative material on the subject matter covered. It is sold with the understanding that the authors and publisher are not engaged in rendering professional services in the book. If the reader requires personal assistance or advice, a competent professional should be consulted.

The authors and publisher specifically disclaim any responsibility for any liability, loss, or risk, personal or otherwise, which is incurred as a consequence, directly or indirectly, of the use and application of any of the contents of this book.

**Publisher:** Marie Butler-Knight
**Product Manager:** Phil Kitchel
**Managing Editor:** Jennifer Chisholm
**Acquisitions Editor:** Mike Sanders
**Development Editor:** Michael Koch
**Senior Production Editor:** Christy Wagner
**Copy Editor:** Jan Zunkel
**Cover Designer:** Doug Wilkins
**Book Designer:** Trina Wurst
**Creative Director:** Robin Lasek
**Indexer:** Lisa Wilson
**Layout/Proofreading:** Angela Calvert, Svetlana Dominguez, John Etchison, Gloria Schurick
**Marketing and Publicity:** Dawn Werk, 317-581-3722

On September 11, 2001, the United States was plunged into a crisis by well-coordinated, well-planned, and well-financed terrorist attacks that destroyed the World Trade Center and severely damaged the Pentagon.

We took these incomprehensible actions very personally, especially the Manhattan tragedy. We are both native New Yorkers; one of us could see the World Trade Center towers being built from his apartment, the other was privileged to visit the construction site and tour the incomplete buildings before their "outer skin" was put on. Both of us had children and grandchildren living within sight of the twin towers and, over the phone, we shared their horror as they witnessed the horrific destruction.

This book is dedicated to our wives, Estelle and Sue, and to our children: Charles and Lawrence Ubell, Anna Ubell Garcia, James and Andrew Merlis and their families. It is also dedicated to our grandchildren: Alexandra Ubell, Mollie Ubell Garcia, and Jasper and Zeke Merlis.

It is our hope that this new generation grows up in a world where no one will ever have to write a book like this; where conservation of vital resources will be not a matter of political debate but a way of life and that their world will be one powered by clean, abundant energy from dependable and renewable sources; a world where corporate avarice is replaced by common sense and where international energy extortion is replaced by worldwide technological cooperation; a world where the word "terrorism" is used only in history classes.

In the aftermath of the September 11 attacks, millions of us asked what we could do. Responses ranged from "give blood" to "give money" to "say a prayer." One recurring theme was "go back to living your normal life."

We would like to suggest an exception to that; to amend that "normal" lifestyle to an energy-saving lifestyle. The more we conserve as individuals, the stronger our nation will be and the more secure our future will be.

# Contents

| | | |
|---|---|---|
| | Introduction | vii |
| 1 | Tracking Your Energy Dollars | 1 |
| | What Is Your E.Q.? | 1 |
| |    Your Energy Quotient Audit | 2 |
| |    How Well Did You Score? | 8 |
| | Do You Know Where Your Energy Dollar Goes? | 9 |
| 2 | Insulating Your Home | 13 |
| | Before You Begin | 14 |
| |    Know Your R-Values | 14 |
| |    Determine Your Insulation Needs | 16 |
| |    Know Your Insulation Options | 18 |
| |    Know Your Local Insulation Regulations | 21 |
| |    Know Where to Insulate | 22 |
| |    Check Your Vapor Barriers | 25 |
| | Installing Insulation | 27 |
| |    Getting the Tools You Need | 28 |
| |    Insulating Your Attic | 30 |
| |    Insulating Walls | 34 |
| |    Insulating Your Unused Rooms and Garage | 40 |
| |    Insulating a New Construction | 41 |
| |    For More Information | 42 |
| 3 | Keeping Out the Cold | 43 |
| | Taking Care of Windows and Doors | 43 |
| |    Check Your Windows | 44 |

|   |   |   |
|---|---|---|
|   | Upgrade to Storm Windows and Doors | 61 |
|   | Weatherstrip Your Doors | 70 |
|   | Shopping for Roofing and Siding Materials | 76 |
|   | Using Smart Landscaping | 76 |
|   | For More Information | 79 |
| 4 | Living Efficiently Year Round | 81 |
|   | How Do You Heat? | 81 |
|   | Turn Down That Thermostat | 83 |
|   | Use a Humidifier | 86 |
|   | Make Your Old Furnace More Efficient | 88 |
|   | Your Fireplace | 104 |
|   | Cool, Man, Cool: Air-Conditioning and Ventilation | 116 |
|   | Tolerate Higher Temperatures | 116 |
|   | Service Your Air Conditioner | 116 |
|   | Be a Fan of a Fan | 121 |
|   | Explore Additional Ventilation Alternatives | 123 |
|   | For More Information | 125 |
| 5 | Stretching Your Energy Dollar—at Home and on the Road | 127 |
|   | Saving Around the House | 127 |
|   | The Codes of the West, East, North, and South | 131 |
|   | Get More Efficiency from Your Old Appliances | 136 |
|   | Let There Be Light—but Only If You Need It | 143 |

| | |
|---|---|
| If You Can Stand the Heat, Stay in the Kitchen | 150 |
| Energy Monitors | 155 |
| Saving in Your Home Office | 156 |
| Saving Behind the Wheel | 158 |
| Gas Price Watch | 164 |
| Buying Economy | 165 |
| Hybrids on the Horizon | 167 |
| Getting a Tax Credit | 167 |
| Photo-Voltaic Panels | 171 |
| Useful Energy Web Sites | 173 |
| For More Information | 177 |

**Appendixes**

| | | |
|---|---|---|
| A | Track Your Energy Expenses | 179 |
| B | Prepare for the Worst | 181 |
| C | For More Information | 187 |
| | Index | 191 |

# Introduction

We were motivated to write this book by soaring energy prices and the prospect of possible electricity, gasoline, and heating oil shortfalls and by our nation's overdependence on foreign sources of energy.

We no longer have the luxury of deciding whether to be energy wastrels or not. We have an obligation to save energy—for our individual good and for the good of the nation.

We all know the shock of high energy prices. What may be less evident is the risk of being dependent on potentially unreliable overseas sources of energy.

During the winter of 2000 and 2001, gasoline prices in the midwest and far-west soared to record levels. People living in those regions felt like victims. During the late winter and spring of 2001, California was struck by a series of rolling electricity blackouts and unprecedented electricity price hikes. Californians felt like victims. Now, with military demand for fuel about to escalate and natural gas supplies not keeping up with demand, the whole nation may soon be feeling like victims.

However, it doesn't have to be that way—you can triumph in this energy cost battle. If you follow just a handful of the recommendations in this book, you will save this volume's purchase price dozens, if not hundreds, of times over. You'll be a winner. If you follow many of the recommendations in this book, you will save thousands of dollars on your future energy bills. You'll be a victor! You'll also be doing your part to preserve energy resources, keep our military well-stocked with fuel, and deprive our enemies of the lifeblood of mass terrorism—your money.

It's that simple—if you win, we all win. If you follow the tips in this book, not only will you save money personally; you'll also help conserve our nation's precious energy resources, keep our environment clean, slow the pace of global warming, and help in the effort to eradicate the enemy. This is a win-win-win situation—*if* you take the steps we recommend in this book.

This book is organized to enable you to go from victim to victor, simply and painlessly.

Before you can save, you have to know where you are spending, so Chapter 1, "Tracking Your Energy Dollars," explains where your energy dollars are going (you may be surprised) and provides a quiz that helps you figure out where you as an individual need the most help. Next, we go through your home and review your lifestyle habits and make recommendations for energy-saving steps, ranging from minor adjustments that cost you nothing to prudent investments that will pay you huge dividends in saved energy dollars *forever!*

Chapter 2, "Insulating Your Home," shows you how you can save a lot of money on your largest single energy expenditure—home heating. With the cost of both natural gas and home heating oil going up, the savings outlined in this chapter will be significant.

Chapter 3, "Keeping Out the Cold," explains how you can make further savings in your home heating (and cooling) bills by making improvements to your doors and windows. Also, we tell you how your landscaping can be designed to save you a lot of money by shading your house in the warmer months and by letting the sun shine through in colder periods.

Chapter 4, "Living Efficiently Year Round," explains the basics of home heating, home cooling, and hot water heating—huge cost items for every homeowner—and gives you a series of steps for saving lots of money on these expenditures.

Chapter 5, "Stretching Your Energy Dollar—at Home and on the Road," provides you with strategies for the most economical operation of your appliances and your automobile. In addition, you'll find lots of extra tidbits called *sidebars* positioned throughout these chapters. These asides are designed to supply you with extra hints, cautions, and money-saving tips, as well as key Web sites you can visit for further information and additional help in saving energy.

The ball is in your court. You can stick with the status quo, shell out big bucks, watch your hard-earned dollars go down the energy drain, and see our dependence on foreign sources of energy remain at dangerously high levels. Or you can join us in the crusade for energy victory. If the last year has shown us anything, it is that the only thing predictable about energy supplies and prices is that the former are going to be spotty, if not sporadic, and the latter are going to climb—sometimes slowly, and sometimes with the speed of a jet fighter.

We believe Americans need not be the hapless—and helpless—victims in the continuing energy drama.

This book will empower you to go from energy victim to energy victor.

## Acknowledgments

No work of this nature springs solely from the minds, resources, and imagination of its authors. We had help from a lot of sources.

We would like to acknowledge the contributions of Estelle Ubell in scrupulously proofreading the manuscript and Lawrence Ubell for his attentive research assistance in marshaling the facts for this volume. Additionally, we would like to thank illustrators Carl Creag and Robert Salanitro and ProGraphics of New York, Inc.

Matthew Barnett rendered invaluable research assistance, Joann Rodriguez aided us with graphics development, and Grantley Thornhill was our mathematics and computer guru, Jules Spodek who meticulously checked our grammar and, of course, Ayhan Turkmen, MEEE, our solar, photo-voltaic expert.

Finally, we'd like to thank Marilyn Allen of the D4EO Literary Agency for her constant support, Mike Sanders of Alpha Books for recognizing the need for a book like this, and Michael Koch for his editing and organizing prowess.

We would also like to thank all of the corporations, societies, and governmental agencies for generously and patiently supplying information not just about their own products and services but about energy conservation in general.

## Trademarks

All terms mentioned in this book that are known to be or are suspected of being trademarks or service marks have been appropriately capitalized. Alpha Books and Pearson Education, Inc., cannot attest to the accuracy of this information. Use of a term in this book should not be regarded as affecting the validity of any trademark or service mark.

# Chapter 1

# Tracking Your Energy Dollars

This book is designed to save you energy and—hence—save you money. On average, 5 percent of Americans' expenditures are for energy! If we could cut that by only 1 percent, think of what we could do with that saved money! Would you rather write a check to the electric utility or to a nice vacation resort? Would you rather pay a gasoline credit-card bill or save a few dollars for your kids' education?

Think you have no choice? *Wrong!* You've got a lot more control than you think you do because, with a few exceptions, most of us are wasting money on energy—a lot of money. We are squandering a finite resource and putting our personal finances and our nation's economy at risk. We can all be a lot more energy efficient and, thus, make our nation and our economy stronger. But before we can save, we have to know how we're spending.

## What Is Your E.Q.?

E.Q.? Energy Quotient! Are you energy-wise or energy-fuelish? Are you a 1,000-hitter or a 100-point energy wastrel? Here's a way to grade yourself. If you reach 1,000 points on the following little energy quotient audit, then you've made just about every improvement you can, and you lead an extremely energy-virtuous existence. But most of us won't score anywhere near 1,000. We've gotten into some really bad habits over the years, and there's lots of room for improvement in most of our homes and in most of our lives. So, take this test, add up your score as you go, and grade yourself, your home, and your lifestyle. And then follow the advice in this book and see how many more points you can gain. You might think of it as a game; a game that can pay you big-dollar prizes as you approach the magic 1,000-point mark and win the GOLD STAR; a game that's really rewarding to play.

## Your Energy Quotient Audit

### A. Heating

| Heating Considerations | Points | Your Score |
|---|---|---|
| **Thermostat Setting (in Fahrenheit)** | | |
| **Winter Day** | | |
| 74 | 0 | _____ |
| 73 | 3 | _____ |
| 72 | 6 | _____ |
| 71 | 9 | _____ |
| 70 | 12 | _____ |
| 69 | 15 | _____ |
| 68 | 18 | _____ |
| 67 | 21 | _____ |
| 66 | 24 | _____ |
| 65 | 27 | _____ |
| 64 | 30 | _____ |
| **Winter Night** | | |
| 65 | 15 | _____ |
| 64 | 18 | _____ |
| 63 | 21 | _____ |
| 62 | 24 | _____ |
| 61 | 27 | _____ |
| 60 | 30 | _____ |

# Chapter 1: Tracking Your Energy Dollars

| Heating Considerations | Points | Your Score |
|---|---|---|
| **Winter Habits** | | |
| You use an electric blanket to allow you to lower the thermostat at night. | 6 | _____ |
| You wear two sweaters indoors so you can lower the thermostat during the day. | 9 | _____ |
| You've installed an automatic flue damper on your heat system. | 30 | _____ |
| Your heating system has been serviced within the last six months. | 15 | _____ |
| You change or clean your heating system filters every month. | 3 | _____ |
| Your heating ducts have no leaks, or the leaks have been taped. | 4 | _____ |
| All heating ducts or steam pipes are insulated. | 4 | _____ |
| Your oil burner burns without smoke or signs of carbon on its surfaces. | 9 | _____ |
| You have a working draft adjuster in your oil or gas burner. | 9 | _____ |
| Your gas burner burns with a clear, blue flame. | 9 | _____ |
| Your gas burner has an electronic ignition system, not a pilot light. | 10 | _____ |
| Your oil burner is a new one with a retention head. | 30 | _____ |
| You use a humidifier during the winter. | 9 | _____ |
| Your radiators or air-supply registers are not blocked by drapes or furniture and are clean. | 27 | _____ |
| You close off rooms not in use, and turn off the heat in them. | 25 | _____ |
| You keep your fireplace dampers shut, or you have glass fireplace doors. | 25 | _____ |
| On winter days, you open your drapes on the south side of the house and close them at night to take advantage of radiant heat from the sun. | 10 | _____ |
| **Water Heater and Usage** | | |
| Your water heater has an insulated jacket. | 15 | _____ |

*continues*

## A. Heating (continued)

| Heating Considerations | Points | Your Score |
|---|---|---|
| **You've Set Your Water Heater Temperature at:** | | |
| 110°F | 12 | _____ |
| 120°F | 9 | _____ |
| 130°F | 6 | _____ |
| 140°F | 3 | _____ |
| 150°F* | 0 | _____ |
| 160°F* | –3 | _____ |
| 170°F* | –6 | _____ |
| 180°F* | –9 | _____ |
| (*Note: Water temperature in excess of 140°F could be hazardous!) | | |
| You drain sediment from the water heater every month. | 5 | _____ |
| You've installed a solar water heater. | 44 | _____ |

## B. Insulation (Including Windows and Doors)

| Insulation Type | Points | Your Score |
|---|---|---|
| **Attic Insulation** | | |
| None | –15 | _____ |
| 2" or R-4 | 0 | _____ |
| 4" or R-11 | 15 | _____ |
| 6" or R-19 | 30 | _____ |
| 8" or R-24 | 45 | _____ |
| 10" or R-30 | 60 | _____ |
| 12" or R-38 | 75 | _____ |
| 14" or R-43 = a layer of R-19 and R-24 | 80 | _____ |
| All your insulation has vapor barriers. | 6 | _____ |
| **Insulation in Your Exterior Walls** | | |
| None | –6 | _____ |
| 3" or R-11 | 25 | _____ |
| 5" or R-15 | 33 | _____ |

| Insulation Type | Points | Your Score |
|---|---|---|
| **Insulation in Crawlspaces** | | |
| None | 0 | _____ |
| 6" or R-19 | 25 | _____ |
| 6" or R-19 with vapor barrier | 35 | _____ |
| Your attic and crawlspace are ventilated. | 15 | _____ |
| There is no basement crawlspace. | 30 | _____ |
| All outlets and switch plates on the exterior walls are insulated. | 6 | _____ |
| Your foundation wall is insulated. | 30 | _____ |
| You have a vapor barrier on the foundation wall. | 9 | _____ |
| You have storm windows. | 25 | _____ |
| You have storm doors or an enclosed vestibule. | 15 | _____ |
| **Windows and Doors** | | |
| Your windows are not drafty. | 30 | _____ |
| Your doors are not drafty. | 15 | _____ |
| Cracks at doors and windows, where wood and masonry meet, are fully caulked. | 30 | _____ |
| All window glass has full putty. | 15 | _____ |
| You have double-glazed windows. | 25 | _____ |
| You have broken windows (deduct for each). | –9 | _____ |

## C. Air Conditioning and Ventilation

| Air Conditioning and Ventilation | Points | Your Score |
|---|---|---|
| **Thermostat Setting (in Fahrenheit)** | | |
| 74 | 0 | _____ |
| 75 | 3 | _____ |
| 76 | 6 | _____ |
| 77 | 9 | _____ |
| 78 | 15 | _____ |
| 79 | 15 | _____ |
| 80 | 18 | _____ |

*continues*

## C. Air Conditioning and Ventilation (continued)

| Air Conditioning and Ventilation | Points | Your Score |
|---|---|---|
| You have no air conditioning. | 20 | ____ |
| You use natural ventilation and wear lightweight clothes in summer. | 5 | ____ |
| You close drapes on hot sunny days. | 5 | ____ |
| You close windows and doors on hottest days. | 5 | ____ |
| Your air conditioners all have 11 to 15 or higher SEER (Seasonal Energy Efficiency Ratios) Central System. | 10 | ____ |
| You have window air-conditioning units and each unit has a 10 or higher EER (Energy Efficiency Ratio). | 5 | ____ |
| Your air-conditioning units are shaded or are on the north side of your house. | 5 | ____ |
| You have and use window or wall units. (Deduct 10 points per unit.) | –10+ | ____ |
| You have an attic ventilation fan. | 15 | ____ |

## D. In and Around the House

| Appliance or Usage | Points | Your Score |
|---|---|---|
| **Lighting** | | |
| You've installed fluorescent lights in the kitchen. | 15 | ____ |
| Your closet lights are on auto-switch or timer switches. | 8 | ____ |
| You've replaced multiple low-watt bulbs with single high-watt bulbs for the same light value but lower total wattage. | 9 | ____ |
| You habitually turn off lights when leaving a room. | 15 | ____ |
| You have energy-saving (solid state) dimmer switches. (Add 5 points per switch.) | 5 | ____ |
| **Television** | | |
| You have an instant-on TV set. (Deduct 5 points per set.) (Note: Older units are more energy efficient.) | –5 | ____ |

| Appliance or Usage | Points | Your Score |
|---|---|---|
| You turn off your TV, stereo, and computer when not in use. | 15 | _____ |
| You fall asleep with your TV on. | –30 | _____ |
| **The Kitchen** | | |
| Your refrigerator is in a cool spot. | 5 | _____ |
| Your refrigerator uses less than 800 Kilowatt-hours per year. | 9 | _____ |
| You close the refrigerator door quickly, rather than dawdle. | 5 | _____ |
| You have a manual-defrost refrigerator. | 10 | _____ |
| You defrost it regularly. | 5 | _____ |
| You have an automatic defrost refrigerator. | –5 | _____ |
| You clean the coils on the back or on the bottom of the refrigerator regularly. | 5 | _____ |
| The refrigerator door gasket fits tightly and is not rotted or loose. | 9 | _____ |
| You air-dry dishes rather than use your dishwasher's drying cycle. | 6 | _____ |
| You have a flow restrictor in your kitchen faucet. | 6 | _____ |
| Your gas range has an electronic ignition system, rather than a pilot light. | 29 | _____ |
| All of your gas burners have a clean, blue flame. | 15 | _____ |
| **The Laundry Room** | | |
| You wash with cold water. | 9 | _____ |
| You often use a clothesline instead of your dryer. | 9 | _____ |
| You turn off the iron when you're not using it. | 9 | _____ |
| You forget to turn off the iron. (Watch out: fire hazard!!!) | –18 | _____ |
| **The Bathroom** | | |
| You shower rather than take baths. | 15 | _____ |
| Your shower has a flow restrictor. | 6 | _____ |
| You fix leaky faucets promptly. | 15 | _____ |
| **Outside the House** | | |
| You have trees and shrubs that are placed to allow sun in winter and block wind in cold weather, but that shade your house in summer. | 40 | _____ |

## E. Bonuses

| Bonus | Points | Your Score |
|---|---|---|
| You've installed a clock thermostat with day and night settings. | 15 | _____ |
| You've installed a clock thermostat with double-setback capability. | 20 | _____ |
| You've installed a heat-producing greenhouse on the south side of the house. | 44 | _____ |
| You've installed an energy monitor. | 45 | _____ |
| You've installed a wood-burning stove (if wood is inexpensive in your area). | 55 | _____ |
| You've installed a windmill, solar heating and/or cooling, or any other renewable energy source. | 150 | _____ |
| **Your total score** | | _____ |
| **Best possible score: 1,350** | | |

Note: The possible total will vary from house to house, from lifestyle to lifestyle, and from one area of the country to another, because points are given for either gas furnace- or oil furnace-heated houses; for torrid, temperate, and frigid zones, which have different heat-loss factors; and for heat gain or loss from season to season.

## How Well Did You Score?

| If Your Score Was | Your E.Q. Rating Is | You Can Save |
|---|---|---|
| 0–99 | Very Poor | 50 to 75 percent |
| 100–199 | Poor | 45 to 70 percent |
| 200–299 | Fairly Poor | 40 to 65 percent |
| 300–399 | Fair | 35 to 60 percent |
| 400–499 | Tin Star | 30 to 55 percent |
| 500–599 | Copper Star | 25 to 50 percent |
| 600–699 | Nickel Star | 20 to 45 percent |
| 700–799 | Bronze Star | 15 to 40 percent |
| 800–899 | Silver Star | 10 to 35 percent |

| If Your Score Was | Your E.Q. Rating Is | You Can Save |
|---|---|---|
| 900–999 | Platinum Star | 5 to 30 percent |
| 1,000 or more ... | Gold Star | 0 to 5 percent |

Chances are, however, that you haven't gotten to 1,000 points yet, so let's begin saving energy and money. But, before you can save money, you have to know where you are spending that money.

## Do You Know Where Your Energy Dollar Goes?

You're paying a lot more for energy of all kinds these days—oil, gas, and electricity cost double (and in some cases, nearly triple) what they ran you just a few short years ago. The higher cost of energy is also reflected in the price of everything else you buy; energy-related price hikes have been shooting out of sight recently. (After all, *everyone's* an energy-consumer, including your grocer, delivery person, food packager, even your government—and *everyone's* going to pass on his or her increased energy costs to you in the form of higher prices.)

There's not much you can do about your supermarket's energy bills. However, on a direct, personal, *home* level, if you know *where* you're spending money, you can draft a plan of attack to conserve and save some of that money. So, *do* you know where your energy dollars are going?

The greatest part of your energy budget goes toward keeping that house of yours warm and comfortable. Heating fuels consume about 57 percent of the average family energy budget.

Another 15 percent pays for the hot water in your shower, bathtub, clothes washer, or kitchen sink.

The remaining 28 percent pays for cooking, lighting, running that toaster, the TV, the computer, the scanner, the fax machine, the vacuum cleaner, the hair dryer, and the dozens of other appliances we all depend on.

## 10  Save Energy, Save Money

*Where your home energy dollar goes.*

Remember, these are *approximate* figures. Obviously, if you live in a warm climate, you spend less to heat your house and more to cool it. And if you're one of those self-reliant types with few (or even no) electrical appliances except your house lights, the proportions change again.

Regardless of the exact dollar figure, however, *all* of us can make some improvements, save some energy, and save some money.

We may have to spend some money now to save in the future, but lower fuel bills and lower utilities costs will very quickly pay back those expenditures. In fact, you should think of any energy-saving improvement in your home as one-time investments virtually guaranteed to pay for themselves many times over! And a lot of savings can be realized without spending a dime ...

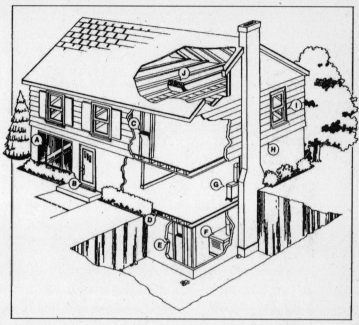

Your house will save you heating dollars if it has the following: (A) storm windows; storm doors; insulated exterior walls; insulated cellar ceilings or crawlspaces; (E) insulated cellar walls; (F) an upgraded, clean furnace; (G) a supplementary wood-burning stove; (H) caulking where the foundation meets the siding; (I) caulked and weather-stripped windows; and (J) an insulated attic.

# Chapter 2

# Insulating Your Home

What do you wear on a cold day? Do you walk outdoors with a thin, threadbare sweater on? If you do, you feel mighty cold. That's because that thin sweater of yours is letting your body heat escape into the cold air.

Now, very few of us walk out on a frigid day dressed like that. Usually we put on something heavier than a skimpy sweater. Maybe we put on a down-filled parka; one of those quilt-like garments can keep us comfortable at very low temperatures. Why? The garment generates no heat. It merely holds in our body heat. In other words, it *insulates* us. How does it do that? By creating air spaces, which prevent our body heat from escaping by blocking or slowing down the heat flow or heat transfer.

If all we've got on is that thin sweater, there is a way we can keep warm. For example, we can jump up and down, run around, and do other vigorous exercise—in other words, we can keep warm by burning calories faster and generating more body heat to keep up with the heat loss. There is only a limit to how long you can keep that up, before you collapse.

Well, the situation's exactly the same with our homes. The house "wears" its insulation for the same reason we wear an outer garment in cold weather—to keep heat in. And the house's equivalent of jumping up and down and running around is to burn more heating fuel—whether it's gas, oil, or electricity. The question is this: What does your house wear: a thin, threadbare sweater or a down-filled parka? If it's wearing that thin sweater,

*What does your house wear in winter?*

then the house will have to burn more energy to keep warm. And it's a lot cheaper these days to buy your house a down-filled parka than it is to feed the heating system more fuel. A down-filled parka for your house is added insulation. And the effectiveness of that parka is measured in R-values.

## Before You Begin

It's important to understand R-values, because then you'll know just how heavy a parka to buy your home.

### Know Your R-Values

Imprinted on every batt or blanket of insulation or on every bag of loose insulation is its R-number. The higher the R-number, the more insulation value you're getting, and the more money you're saving. (Maybe instead of rating insulating material R-3 and R-4 and R-5, it would be more graphic to rate it with dollar signs: $-3, $-4, $-5.)

The "R" in R-numbers is not just an arbitrary letter. It stands for "resistance"—resistance to heat loss in cold weather; and, in hot weather, resistance to the sun's heat.

How can insulation work both ways? Because heat flows naturally to cooler areas. Thus, in the winter, the heat flows from your nice, warm house to the frigid outdoors, and in the summer the heat flows from outdoors toward your cooler house. Insulation blocks or resists that flow, helping your house stay warmer in winter and cooler in summer.

# Chapter 2: Insulating Your Home 15

Proper insulation can save you 20 to 40 percent of your home-heating bill in the winter and about 10 to 15 percent of your cost for home cooling in the summer.

How many R's do you need? Well, that all depends on where you live. Find your home-heating zone on the following map.

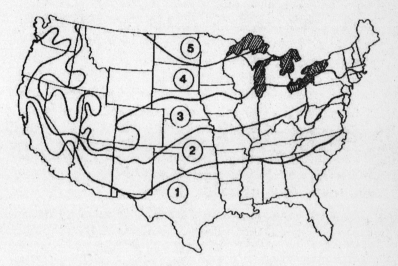

*Heating zones, continental United States.*

Now, check the following table to see how much insulation you need in your attic floor and other areas.

## Recommended R-Values

| Heating Zone | Attic Floors | Ceilings over Unheated Crawlspace or Basement |
|---|---|---|
| 1 | R-38 | R-22 |
| 2 | R-38 | R-22 |
| 3 | R-49* | R-22 |
| 4 | R-49* | R-22 |
| 5 | R-49* | R-22 |

*Note: R-49 can be achieved by using one layer of R-38 plus a layer of R-11, which has a thickness of about 18 1/2 inches.

For a full standard 2×4 stud wall, insulation can only be 3¹/₂ inches thick, and depending on material used, the R-value can range from R-11 to R-13. If the wall thickness is 5¹/₄ inches thick for a 2×6 stud wall, the range can be R-13 to R-22. This should be the code for all American housing. Some newer homes are now using this standard.

For insulation of exterior walls, the R-value depends on the material used and the thickness of the wall. For example, for a full standard 2×4 stud wall, insulation can only be 3¹/₂ inches thick, and depending on material used can range from R-11 to R-13. If your exterior wall is 5¹/₄ inches thick (as is the case with a 2×6 stud wall), the range is R-13 to R-22. This should be the code for all American housing. Some newer homes are now using this standard.

## Determine Your Insulation Needs

If your house has no attic (most flat-roof houses don't), chances are there's a cockloft, a space between roof and ceiling, at the top-floor level. Cocklofts can—and should—be insulated. How do you know if your cockloft has enough insulation? Inspect it. Some cocklofts have hatches that are easily removed. If yours doesn't, remove a ceiling light fixture and check around the edges for insulation. Or, in a closet, poke a hole in the ceiling material and check. (Cocklofts are difficult to insulate. A skilled and reputable professional should do that work.)

*Measure your present attic insulation with a ruler.*

How do you check to see if your exterior walls (those facing the outdoors) are insulated? You have four options:

1. Remove a light-switch plate or electrical-outlet plate on an exterior wall and check around the outside edges of the terminal box with a flashlight.
2. Go into your basement and look up between the exterior wall's 2×4s if the framing is open in the basement (you'll find this is the case in unfinished basements).
3. Go into a closet against an exterior wall, poke a hole in the wallboard with a screwdriver, and examine it with a flashlight.
4. Look for "ghosts": On an exterior wall or a ceiling that hasn't been painted for a while, dust particles accumulate in vertical strips between the wall and framing, making a ghostly outline of the studs on the wall (as shown in the following figure). This occurs *only* on uninsulated walls and ceilings.

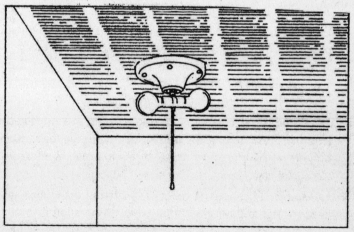

"*Ghosts*" *on an uninsulated ceiling.*

To see if your floors are insulated, check in your basement (if unfinished) or crawlspace. If you've got a finished basement with a ceiling, remove the light fixture and check with a flashlight.

Now that you know *where* you need insulation, you're ready to go. But wait! What *kind* of insulation do you need?

## Know Your Insulation Options

There are three basic types of insulation; all are highly effective, if properly installed:

- **Inorganic.** Products made from minerals, such as fiberglass, fibrous stone, and vermiculite.
- **Organic.** Products made from vegetation, such as wood, seaweed, and waste-paper products.
- **Chemical.** Products made from petroleum.

Within each basic type are insulating materials of different characteristics.

There are five ways of applying insulation:

- **Batts.** These are rolls of insulation material, which are placed within the framing of a structure. Batts are easy to apply—often needing only to be rolled out in your attic space or stapled between the studs of a wall. Almost any do-it-yourselfer can install batts of insulation.
- **Blankets.** Similar to batts, blankets come in uniform-cut lengths. They are used in similar areas and, like batts, they are easy to handle and present little challenge to most do-it-yourselfers.
- **Loose fill.** These are particles of insulating materials packaged in bags and are easily poured between wood frames in your attic. Again, this is an easy-to-install material for most people. Or, professionals using specialized equipment can blow loose insulation into place.

- **Foam.** These are chemicals that are blown or pumped into the area to be insulated. They then solidify, forming a barrier to heat loss. Only skilled, trained professionals should apply foam insulation.
- **Boards.** This is insulation material that resembles ceiling tiles. It can be cut with a razor knife or saw, and then nailed, stapled, or glued into position.

| | Nonorganic (Mineral) | | | | | Chemical | | | Organic |
| --- | --- | --- | --- | --- | --- | --- | --- | --- | --- |
| R-Value | Fiberglass, Batts, and Blankets | Fiberglass, loose, blown, or poured | Rock-wool, Batts, and Blankets | Rock-wool, loose, blown, or poured | Vermiculite loose, blown, or poured | Perlite loose, blown, or poured | Plastic foam boards | Urethane boards | Ureafoam pumped | Cellulose fiber loose, blown, or poured |
| R-11 | 3½–4¼" | 5" | 3" | 4" | 5" | 4" | 2¼" | 1½" | 2" | 3" |
| R-19 | 6–6¼" | 8–9" | 5" | 6–7" | 9" | 7" | 4¼" | 2¾" | 3½" | 6" |
| R-22 | 6½–9" | 10" | 6" | 7–8" | 10½" | 8" | 4½" | 3" | 4" | 6" |
| R-30 | 9½–11½" | 13–14" | 9" | 10–11" | 14" | 11" | 6½" | 4½" | 5½" | 8" |
| R-38 | 12–13" | 17–18" | 10½" | 13–14" | 18" | 14" | 8½" | 5½" | 7½" | 10" |
| R-49 | 15–21 | 22–23" | 13" | 17–18" | 23" | 18" | 11" | 7" | 10" | 13" |
| Fire Resistant | Yes | Yes | Yes | Yes | Yes | Yes | No | No | Yes | No, if not treated |
| Moisture Resistant | In some cases | In some cases | No | No | In some forms | In some forms | Yes | Yes | Yes | No |
| Vermin Proof | Yes | Yes | Yes | Yes | Yes | Yes | Yes | Yes | Yes | No, if treated improperly |
| Rot Proof | Yes | Yes | Yes | Yes | Yes | Yes | Yes | Yes | Yes | No, if treated improperly |
| Corrosive | No | No | No | No | No | No | No | No | Yes, in some cases | Yes, if treated improperly |

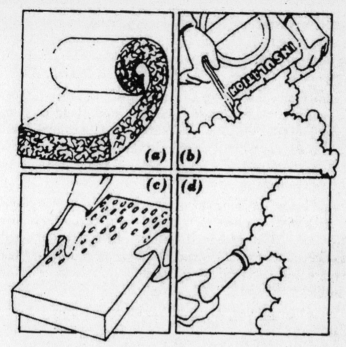

*Four types of insulation: (a) batts, (b) loose fill, (c) boards, and (d) foam.*

## Know Your Local Insulation Regulations

Before you buy any insulation, you should know that different cities, towns, counties, and states have different flammability standards. Before you spend your hard-earned money on any insulation, check your local fire and building ordinances to see which types of materials are acceptable in your area. Chances are good that these regulations are posted on your municipality's Web site or your local fire department's or building department's Web site.

Beware of insulation that is not wrapped in factory packaging or that bears the names of unfamiliar firms. Some fly-by-night operators are cashing in on the rush to conserve energy by making improperly treated and even dangerous insulation materials, and selling them at cut-rate prices.

You may pay more for a well-known brand name, but the added cost is an investment in your family's safety and your own peace of mind. Remember, insulation is a one-time investment that pays dividends forever, so don't cut corners and take risks.

Insulation should have a label from Underwriters Laboratories and the National Association of Home Builders. Also, look for the U.S. Department of Energy and Environmental Protection Agency's ENERGY STAR label. Products with the ENERGY STAR meet or exceed strict government efficiency and performance standards.

And before installing any of the chemical insulation materials, check with your local office of the United States Consumer Product Safety Commission. There have been many reported cases of chemical insulation materials giving off toxic fumes that have caused eye, nose, and throat irritations, and many are highly flammable.

Now that you've graduated from Insulation University, it's time to learn.

## Know Where to Insulate

Remember that down-filled parka you were going to put on your house? Well, in considering *where* you need insulation, let's think about the parka again.

If you're outside on a cold day with a parka that's a couple of sizes too large for you, you're going to feel colder than if you're wearing one that fits you properly. That's because a large, loose parka will let cold air in and your body's heat out.

Also, you can have the best-insulated, best-fitting parka in the world, but if you leave the zipper unzipped, it's not going to do you much good.

It's much the same way with your house. You want to keep the insulation close to the heated areas (you don't want to keep warm air in your attic; you want to keep it in the living quarters *beneath* the attic, so you insulate the attic floor, not the attic ceiling or between the rafters), and you don't want to leave "unzipped" gaps through which the warm air will escape (such as an uninsulated attic trapdoor).

### Caution

Urea Formaldehyde (UF) foam insulation (which is pumped into walls only by contractors) has been installed in half a million or more American homes. The Consumer Product Safety Commission reports hundreds of consumer complaints about unhealthful effects from formaldehyde gas released by UF foam insulation. The complaints include difficulty in breathing, eye and skin irritations, headaches, dizziness, nausea, vomiting, and severe nosebleeds. The commission is also studying possible long-term effects of formaldehyde gas exposure, such as birth defects and links to cancer. UF foam, while largely banned, is still used by some unscrupulous contractors. Insist that your contractor show you the labels on any insulation materials that he or she pumps or blows into your exterior walls.

The following illustration shows you where insulation should be placed.

When insulating your home, you should pay close attention to the following areas:

- **Ceilings.** Especially ceilings with cold spaces above.
- **Exterior walls.** This includes the short walls of a split-level house, which are often neglected. You should also insulate walls between living space and unheated garages or storage rooms. If you have open framing in your garage, you can do the job yourself, but with garage walls that are enclosed on both sides by plasterboard (Sheetrock), you should hire an insulation contractor to do the job. The contractor will have to blow the insulating material into the wall cavity—not a job for a do-it-yourselfer.

◆ **Floors above cold spaces.** Crawlspaces, garages, open porches, and any portion of a floor in a room that extends beyond the wall below.

The following figure illustrates the areas, nooks, and crannies of your home that should be insulated.

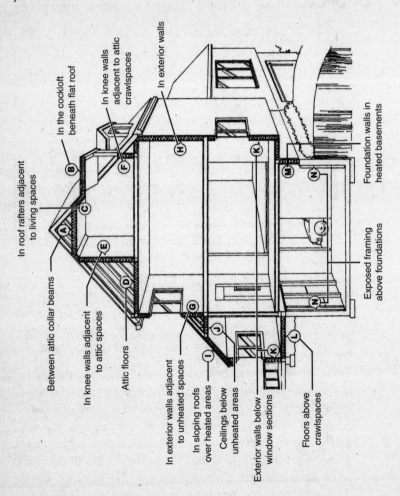

Don't let the previous figure overwhelm you! There are lots of places to install insulation, but in almost every case it's an easy job; you'll need only minimal mechanical skills.

Think about this: Most insulation is hidden from view, so you don't have to install it with a cabinet-maker's precision. Because of this, insulating is kind of fun—it's so much less demanding than a lot of other household repair jobs. The chances are good that you've got everything you need for the job around your house, except the insulating material. So, you've got no excuse—get on with it and watch those fuel savings pile up!

## Check Your Vapor Barriers

You know *where* to insulate, but there's still one more thing you need to know about—vapor barriers.

Well-insulated attics, crawlspaces, storage areas, and other closed cavities need to be well ventilated to prevent excessive moisture build-up.

It doesn't rain inside your house, but it does dew.

Walk out on your lawn on a cool summer morning and you'll find the ground wet—not with rain, but with dew.

Pour yourself a glass of ice water in a nice, warm house and you'll find the outside of the glass wet—from dew.

Install insulation without a vapor barrier and dew will form, which will rot wood, corrode electrical wiring, and erode plaster and wall board.

If it doesn't rain in your house, why do you have dew?

Warm air holds more moisture than cold air does. Your breathing adds moisture to the air. Cooking increases the humidity inside the house. And so does a humidifier. When that warm, moisture-laden air hits the cold surfaces in unheated portions of the attic and walls, dew forms. But if you can place a moisture-blocking barrier between the cold surface and the heated air, dew won't form. Unfortunately, insulation won't do that job by itself—in fact, most insulation materials would collect the moisture, and wet insulation doesn't work. So, you need a *vapor barrier* between the warm portions of the house and the insulation.

*In walls where there is no vapor barrier, dew forms when heated interior air hits colder outside air between walls.*

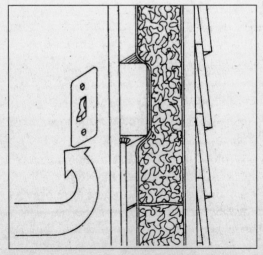

*Vapor barrier blocks moisture in heated air and keeps it in the house, preventing dew from forming.*

Many bales and blankets of insulation have a vapor barrier already attached. All you have to remember is to install the insulation with the vapor barrier *closest* to the heated area (Don't panic. You won't have to hold this book in one hand while you install the insulation with the other. The instructions on how you place the vapor barrier are printed right on the vapor barrier.) For other forms of insulation, it is necessary to lay down or nail up a vapor barrier (usually aluminum foil, saturated kraft paper, or plastic sheeting) before installing the insulation.

Often, when insulating a cockloft or exterior wall, it is impractical or impossible to install a vapor barrier. In those cases, you can create a moisture-blocking effect by papering the wall with vinyl wallpaper or by painting the wall or ceiling with two coats of good-quality oil-based or metallic paint.

Okay, you're ready to begin ...

## Installing Insulation

It's so easy, you'll feel like a million bucks doing it. And you'll save hundreds, eventually thousands, of dollars. And as a bonus for insulating your home, you'll find your environment much quieter. Just as insulation makes it hard for heated air to escape your walls, it also makes it hard for outdoor noises—traffic, wind, and the like—to penetrate your walls.

To figure out how many square feet of insulation you'll need, measure the length and width of the area to be covered, multiply the two dimensions, and you'll have the square footage of insulation material. Now measure the distance between the wall studs or joists (those wooden frames). If you're using batts or blankets of insulation, you'll want to buy them in the right widths (batts and blankets are made in 16-, 20-, and 24-inch widths). If you're using a poured insulation material, the label will tell you how many square feet each bag of material will cover at each of several R-numbers.

## Loose or Poured Fiberglass Insulation

| R-Value | Minimum Weight | Minimum Thickness |
|---|---|---|
| To obtain a thermal resistance of | Weight of installed insulation should not be less than (in lb/sq. ft.): | Should not be less than (in inches): |
| R-11 | 0.179 | 4.75 |
| R-13 | 0.209 | 5.50 |
| R-19 | 0.301 | 7.75 |
| R-22 | 0.353 | 9.00 |
| R-26 | 0.418 | 10.50 |
| R-30 | 0.485 | 12.00 |
| R-38 | 0.615 | 14.75 |
| R-44 | 0.712 | 16.75 |
| R-49 | 0.800 | 18.50 |
| R-60 | 0.986 | 22.00 |

If you're in doubt, buy a little extra to be on the safe side—just make sure you can return unopened bags and still-wrapped batts and blankets and get a refund.

### Getting the Tools You Need

The following figure shows the tools and equipment you need for insulating jobs.

### Caution

When you're working in your attic, watch out for nails protruding through roofing boards. You want a warm house, not a hole in the head. Don't step off the floor framing or walking board. You don't want a hole in your ceiling below, either.

## Chapter 2: Insulating Your Home

A   Duct or masking tape, $1\frac{1}{2}$ to 3 inches wide
B   Hammer
C   Serrated-edge kitchen knife
D   Dust mask
E   Assorted nails
F   Scissors or shears
G   Gloves
H   Lighting (extension light for those dark corners)
I   Tape measure or ruler
J   Rake (if using loose insulation)
K   Walking board ($\frac{3}{4}$-inch thick, 16 inches wide, 4 feet long)
L   Heavy-duty staple gun and staples
M   Long-sleeved shirt with buttoned wrist cuffs

*The tools and equipment you'll need for insulating your home.*

Be sure to protect yourself before beginning your insulation project. You need the long-sleeved shirt, the gloves, and the dust mask to protect you from the fibers in the insulation and the dust in the attic. Never install insulation without them. You need to be aware that some people develop reactions to certain insulating materials—reactions such as skin irritations, burning eyes, and sore throats. Check with your doctor if you develop these symptoms.

Now you're ready for an easy and rewarding job.

## Insulating Your Attic

One of the first rules of winter survival is: Wear a hat. We humans lose a lot of body heat through an uncovered head and ears. Well, it's the same thing with our houses. An uninsulated or underinsulated attic is the same as going out bare-headed on a freezing day. So start insulating your home from the top. Attic insulation is not only the most important insulating job you can do; generally it's the easiest and least expensive, too.

*When insulating an attic floor with batts, be sure you wear that dust mask!*

*Before installing loose insulation, be sure to staple in a vapor barrier. Then, pour in the insulation and level it off.*

### Starting from Scratch

If there is no insulation in your attic, roll batts of

insulating material between the joists (frames) with the vapor barrier down, facing the living space below.

Alternatively, you can also place vapor barriers between joists; then, pour in loose insulating material and level with a rake or straight stick.

If you end up tearing the vapor barrier, don't throw it away—patch it! Use masking tape and fix vapor barrier tears—no matter how small they are.

## Upgrading the Existing Insulation

If your attic already has some insulation, but needs more, lay new batts of insulation over existing material, at right angles to the joists. But be sure the new insulation you're using has *no* vapor barriers, or slash the vapor barriers on the new batts. This will allow the old insulation to "breathe." Install the new batts with the slashed barrier down.

And if your old insulation has a kraft-paper covering facing up, cut slashes in that, to allow moisture to escape into the attic.

Whether you're installing the first insulation in your attic or increasing the insulation already there, be sure you don't block ventilation openings in the eaves and in gable ends.

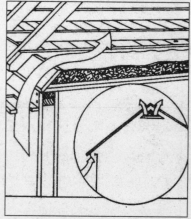

*Your attic must breathe. Cut kraft-paper covering over old insulation before installing new batts. Don't let insulation block eaves. Air must flow up through eaves and out ventilator openings.*

## ⚡ Caution

Don't smoke while working in the attic. The dust could be combustible and cause a fire. Besides, you're supposed to be wearing a dust mask, and you can't smoke with a dust mask on. (Hey, maybe wearing a dust mask all the time can help you quit smoking altogether!)

### Insulating Attic Trapdoors

Remember that unzipped down-filled parka? Well, your attic trapdoor can be like that open parka zipper. Don't forget to insulate it. You can staple insulation material directly to it. (Remember to keep the vapor barrier facing the heated area of the house.)

*Don't forget to insulate your attic trapdoor.*

## ⚡ Caution

Don't insulate around lighting fixtures, motors, or any other electrical equipment mounted through the attic floor (typically, the lighting fixtures and bathroom exhaust fan motors recessed into your ceiling protrude into the attic through the floor). Keep insulation at least three inches from these objects.

Exercise extreme care in handling electrical cables and wires in the attic or anywhere else you are installing insulation. Rough handling could cause a short circuit or a fire.

You can make a frame to keep insulation a safe distance from electrical equipment with sheet metal or a large coffee can.

### Insulating Attic Walkways

That flooring over the attic joists may hide an uninsulated area. Even if you have to temporarily remove the flooring, it's worthwhile to get insulation between those joists. You may be able to squeeze it there by pushing and pulling if the walkway isn't too wide.

*Protect protruding lighting fixtures from insulation with a shield made from sheet metal or from a coffee can.*

*Don't forget to insulate under floorboards or catwalks in your attic.*

### Hint

Insulate all heating and cooling duct-work in attic, basement, or crawlspace with at least two or three inches of fiberglass—to keep the heat in the ducts and flowing to your home's living areas.

*Be sure you insulate around chimneys, ducts, and pipes in your attic.*

## Cost and Savings

### What Will It Cost?

Insulating your attic will cost about 38 cents a square foot to bring your insulation level to R-38 or approximately 12 inches of fiberglass. If you hire a contractor to blow in loose insulation, it will cost you $1.50 a square foot.

### How Much Will I Save?

If you start with an uninsulated attic and you do all we recommend, you can save yourself as much as 20 percent on your home heating bill every year from now on. This is an improvement that can pay for itself in less than one heating season. And it's easier than falling off a log—so what are you waiting for? Do it!

### Insulating Attic Pipes

Vent pipes and electrical conduits and chimneys coming up through your attic floor should have insulation hand-packed around them to prevent warm-air loss through ceiling cracks.

*Insulate all heating and air-conditioning pipes and ducts in crawlspaces and attics.*

### Insulating Walls

Now that your house has a nice, warm hat (attic insulation), it's time to put on its down-filled jacket. You do that by insulating the walls. Wall insulation can be tricky and may require the services of a professional. But don't worry. Insulation will pay for itself, even if you have to hire someone to install it.

### Insulating Finished Exterior Walls

It is nearly impossible for a do-it-yourselfer to insulate a finished exterior wall without tearing off the wallboards. Competent professionals can do this work for you. However, there are some sharp, fly-by-night operators in this industry, and you should exercise extreme caution before you hire someone to blow insulation into your finished walls.

If your state licenses contractors, be sure you are dealing with a licensed contractor. In licensing states, check your contractor's license number against the licensing agency's complaint records (often available on the Internet). In all cases—whether your state licenses or not—check with your local Better Business Bureau to see if complaints have been lodged against the contractor.

> ### Hint
>
> **Energy-Saving Seating**
>
> Whenever possible, place chairs, couches, and beds *away* from exterior walls. You'll feel less of the outside chill so you won't be pushing that thermostat up to increase your comfort.

### Insulating Unfinished Walls

Using blankets the width of your wall studs, insert the blanket between the studs and staple to the studs. (The job is easier if you start from the top and work down, just as you do when you wallpaper your home. What? You don't wallpaper your home? Well, after you master insulating from the top down, you can begin.) Be certain the insulation fits snugly against the top piece of framing.

You can also insulate unfinished walls using blankets without vapor barriers; a vapor barrier of two-mil-or-more plastic sheeting or foil-backed gypsum board can then be stapled to the studs. (Remember to keep the plastic taut as you staple it in place.)

## Caution

The vapor barrier must face the *heated* side of the house's interior space and that's an easy job with unfinished walls. But what about a vapor barrier on finished walls? In that instance, you're going to have to rely on an interior layer of vinyl wall covering or a couple of coats of oil-based paint on the living side of the wall.

*Insulating an exposed exterior wall: Staple insulation between studs ...*

*... and then cover with vapor barrier.*

 ## Hint

A lot of heat can escape through light-switch plates and wall sockets on exterior (outdoor-facing) walls. To combat this heat loss, buy precut foam insulation kits at your hardware or building-supply store. These small insulation rectangles are easy to install and pay for themselves very quickly.

It will cost you less than a dollar per outlet or switch plate, and can save you 1.5 percent of your home heating cost.

Be sure to buy only those kits listed by Underwriters Laboratories (UL).

### Insulating Unfinished Masonry Walls

You may have unfinished masonry walls in your basement. To insulate them, you've first got to create wall studs to hold the insulation material. Using masonry nails, fasten 2×2 furring strips in place vertically, placing them 16 or 24 inches apart (measure from the center of each strip to the center of the next strip). Then insulate as with any other unfinished wall.

## Caution

When applying insulation around water pipes, be sure you place the insulating material between pipes and the *outside* wall (the pipes should be exposed, not covered). This will keep the pipes from freezing and bursting.

*Be sure insulation is between pipes and conduits and cold areas, not between them and heated areas.*

### Insulating Door and Window Frames

Be sure you get insulation in the spaces around unfinished walls, doors, and windows. Because the framing area is narrow, loose insulation must be hand-packed into it.

## Cost and Savings

**What Will It Cost?**

To insulate your exterior walls will cost you about $1.50 to $2.00 a square foot.

**How Much Will I Save?**

If your exterior walls are uninsulated, installing insulation—or having it installed by a qualified professional—can save you 15 percent of your home heating bill every year from now on!

*Insulate openings around window and door frames.*

## Insulating the Floor Above a Cold Crawlspace

For this one, you're probably going to have to work flat on your back. But console yourself; that's the way Michelangelo painted the ceiling of the Sistine Chapel.

To begin, drive nails into the floor joists, about 18 inches apart. This will be used later to "lace" in your insulation.

Now shove batts of insulation up between the joists. They should fit snugly enough to remain there, defying gravity, for a short time. While it's still in its gravity-defying mode, lace wire back and forth between the nails to fix the insulation permanently in place. Or you

### Cost and Savings

**What Will It Cost?**

It will cost you 35 cents a square foot to insulate your crawl space.

**How Much Will I Save?**

You can save 10 percent of your home heating bill by insulating your crawl space. And once you get used to working on your back, it's easy.

can nail or staple chicken wire to the joists to hold the insulation (although that's a bigger job).

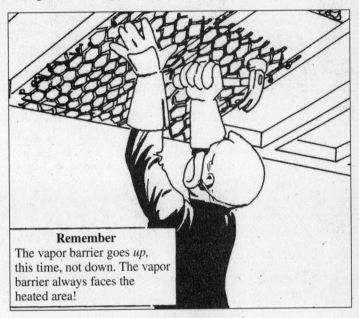

**Remember**
The vapor barrier goes *up*, this time, not down. The vapor barrier always faces the heated area!

*"Lace" insulation in place on crawlspace ceilings with wire. Or nail up chicken wire after you insulate.*

## 40  Save Energy, Save Money

 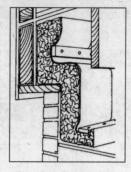 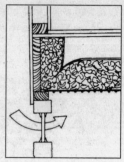

When floor joists are at right angles to the building foundation, nail insulation to the sill.

When floor joist is parallel to the foundation, you'll need a strip of furring to nail the insulation in place.

Be sure that insulation doesn't block crawlspace ventilation grilles.

### Hint

When insulating a crawl space, be certain to insulate the rim joist and the floor joists where the subfloor meets the building end. This will close the last open zipper on your home's down-filled parka. However, be certain that you don't cover foundation vents with insulation.

### Insulating Your Unused Rooms and Garage

If you've got a big, old, partially occupied house, you may want to save on your fuel bills by not heating (or air conditioning) unused rooms. However, when you begin shutting off radiators, convectors, or air vents, remember that you'll save even more by putting some insulation between the rooms you shut off and the rooms you're continuing to heat.

You can do that by insulating the interior walls of the room you're shutting off or—as a quick-fix solution—by hanging blankets or quilts on those walls. Also, if you store a lot of furniture, paperwork, merchandise, or other nonperishables in the unheated room, their bulk will create insulation and cut down heat loss from the parts of the house you want to keep warm.

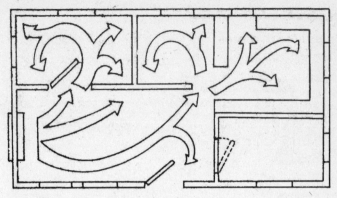
*Block off unused rooms and don't heat them.*

Whether your attached garage is heated or not, it should be sealed up against the winter weather. Weatherstrip and caulk the doors and windows, and insulate the walls that adjoin the living space of the house. (For more details, see the sections on weatherstripping and caulking in Chapter 3, "Keeping Out the Cold.") Remember to keep the vapor barrier facing the living areas, not the garage areas. You might also consider installing a storm door between garage and house. But don't make the garage airtight. Exhaust fumes must be able to escape!

## Insulating a New Construction

If you're planning any new construction, whether it is an addition to an existing home or the construction of a new home, spend the extra money to have 2×6 exterior wall studs instead of 2×4 studs. The 2×6 studs will accommodate two more inches of insulation, enabling you to boost the R value from R-11 to R-19, which is almost double! It will cost you about an additional $10 more per linear foot of building for the larger studs, and the added insulation is only about 13 cents a square foot. It will pay for the construction increase in heating savings in just a few years and many, many times over, during the life of the house.

## For More Information

For further information check the following Internet resources:

| | |
|---|---|
| U.S. Department of Energy | www.doe.gov |
| Energy Star | www.energystar.gov |
| Certainteed (Insulation Mfg.) | www.containteed.com |
| Cellulose Insulation Manufacturers Association | www.cellulose.org |
| Energy Efficiency and Renewable Energy Clearinghouse | www.ornl.gov |
| Insulation Contractors Association of America | www.insulate.org |
| National Association of Home Builders | www.nahb.com |
| North American Insulation Manufacturers Association | www.naima.org |
| Owens Corning | www.owenscorning.com |

# Chapter 3

# Keeping Out the Cold

All your hard work insulating your home will pay big dividends. But to maximize those dividends, you need to weatherproof your doors and windows, too. Having doors and windows that let cold air in during the winter is a little like wearing the warmest down-filled jacket you can find and then leaving it unzipped. Because they can't be insulated, doors and windows are going to be cold spots in the winter and hot spots in the summer. Let's see how you can minimize those chinks in your house's weather armor. Also in this chapter, you'll see how your choices in roofing material, siding and even landscaping can save you heating and cooling dollars.

## Taking Care of Windows and Doors

Do you know where the word "window" comes from? It's a combination of "wind" and "door," and originally meant a door through which the wind blew—literally, the wind's door. In this day of expensive home heating fuels, the last thing you want is for your windows to be open doors for the cold wind. You can take steps to significantly cut heat loss through your windows. Additionally, your windows are a solar heating system.

A window can actually serve your heating needs. It can act as a conduit for the sun's energy and let in radiant warmth faster than it lets out your furnace-created heat. This is the simplest form of solar energy—a form we can all take advantage of, without expensive solar collectors or heat distribution systems. It's called "passive" solar heating.

In general, south-facing windows admit the most heat, followed by east and west exposures. Because north-facing windows admit little heat and tend to conduct room heat *out* of the house, keep blinds, curtains, or shutters drawn on those windows during the colder months. (And remember to do the same for your heat-collecting windows during the night, too.)

To be economically significant, that solar collector window of yours has got to bring in more heat than it loses. Unfortunately, a lot of windows let out more heat than they take in. In fact, in the average house, each improperly insulated window will waste about 1 percent of your home heating dollars. Properly insulating 10 windows can save you 10 percent of your fuel costs.

You can save a lot of money using storm windows, but, if your primary windows are in poor condition, the storm windows won't be as effective. So, the first step to saving money is ... clean that glass. Your passive solar collector (your window) works better clean than dirty. Dirt on the windowpanes absorbs the sun's rays and doesn't let the heat in. Keep those south-, east-, and west-facing windows clean, (and while you're at it, you might as well clean the north windows, too). A fringe benefit: More sunlight means a brighter room, which means you don't have to turn on your lights as much, saving you money on your electric bill.

## Check Your Windows

Check to see if your windows are open doors for the wind. Look for the following:

- ◆ Broken or cracked glass
- ◆ Broken, loose, or missing latches
- ◆ Loose or broken sashes and frames
- ◆ Cracked or missing putty around panes, or damaged weatherstripping
- ◆ Lack of weatherstripping
- ◆ Lack of caulking

## Chapter 3: Keeping Out the Cold

*Inspect windows for broken glass, broken latches, missing caulking, and missing putty.*

### ⚡ Caution

Many pamphlets and books—including some from government agencies—advise testing for air leaks around windows by moving a lighted candle around the frame to see if it flickers. That test can be dangerous. The candle can ignite curtains or draperies. We always use a single sheet of dangling tissue paper around the frame. If the tissue is blown away from the window or sucked toward it, you've got an air leak that's costing you money.

The following sections discuss how to fix each window problem.

## Fixing Broken Windowpanes

To fix a broken windowpane, you'll need the following tools:
- Gloves
- A large piece of cardboard
- Scissors
- A putty knife
- A paintbrush
- An electric iron or propane torch

You'll also need these materials:
- A box of glazier's points
- A can of putty or glazing compound
- Some boiled linseed oil
- Cornstarch
- A piece of glass from the hardware store or home improvement center

To fix a broken windowpane, follow these steps:
1. Remove all the old broken glass from the frame. (Be sure to wear gloves to protect your hands from the jagged edges.)
2. Using the propane torch, heat up the old putty and remove it with the putty knife and pliers. (You can use an electric iron to heat the putty if you don't have a propane torch and don't want to buy one.)
3. Using the pliers, carefully pull out the old glazier's points (those are the little metal triangles or staples that hold the glass in place).
4. Cut your cardboard to fit the window frame, and then trim $1/8$-inch from the top and from one side of the cardboard (this creates a $1/16$-inch margin on each of the cardboard's four sides).
5. Now have a glazier or hardware store cut a piece of window glass to the exact size of the cardboard.

6. Returning to your window, sprinkle cornstarch on your hands and on any surface where you'll be rolling putty. This keeps the putty from sticking where you don't want it.
7. Roll all the putty in the can into a $3/8$- to $1/2$-inch diameter rope.
8. Brush a light coat of linseed oil onto the frame and press the putty rope into the oiled frame (cut the putty rope into six-inch lengths for easier handling).
9. Press the glass into the frame and remove all but $1/16$ inch of putty squeezed out by the pressure of the glass.
10. Wedge the glazier's points into the frame with the blade of the putty knife, spacing them about eight inches apart.
11. Cut the remainder of your putty rope into six-inch lengths and squeeze into the corner of the window frame.
12. Smooth and remove the excess putty with the putty knife, and you've sealed a window and saved some heating dollars.

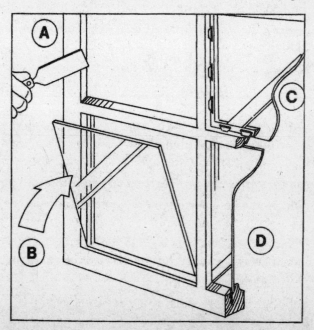

## Fixing Broken, Loose, or Missing Window Latches

That window latch is there for more than security—it presses the top and bottom sashes together to eliminate outside air infiltration or inside heat loss. If the latch is broken, loose, or missing, you're losing heating dollars. To install a new latch, you'll need the following tools:

- Ruler
- Hammer
- Center-punch or awl
- Cake of soft soap
- Medium flat-blade screwdriver

You'll also need these materials:

- Center-latch, also called a sash lock
- Petroleum jelly

To fix broken, loose, or missing window latches, follow these steps:

1. Remove the old, broken, or loose latch, and then close the window tightly.
2. Measure and mark the center of the joined sash frames.
3. Place the turning-latch section of the lock on the center mark at the top of the lower sash frame.
4. With a hammer and awl, punch guide holes through the screw ports on the turning-latch section.
5. Lubricate wood screws by rubbing them over the bar of soap, and then screw the turning latch section firmly into position.
6. Line up the hook section on the bottom of the frame of the upper sash, making sure it is even with the turning-latch section.
7. Punch holes as you did for the first part of the latch.
8. Soap and drive screws into place.
9. Lubricate your new latch with petroleum jelly.

*Window latches keep sashes firmly closed against the cold.*

## Fixing Loose Sashes

You usually know if your sashes are loose because they tell you. No, they don't speak; they just rattle noisily when the wind blows. That rattle means cold air's coming in and warm air's going out. To fix a loose sash, you'll need the following tools:

- Pencil
- Shears
- Ruler
- Small hammer

You'll also need these materials:

- Pad of paper
- A piece of linoleum
- Carpet tacks
- Paraffin wax or silicon spray

To fix a loose sash, follow these steps:

1. If your lower sash is loose, raise it to the top and measure the opening, then add six inches. Write down that measurement.
2. Measure the thickness of the sash and subtract $1/8$ inch. Write down that measurement.
3. Cut strips of linoleum to those measurements and close the window.
4. Slip one strip of linoleum into the sash groove (if your window has metal weatherstripping, you must slip the linoleum under the weatherstripping).
5. Tack the linoleum in place with the carpet tacks and open the window.
6. Tack the six-inch section of linoleum below the top sash frame. Close the window.
7. If the sash still rattles, add more linoleum strips. Lubricate the linoleum strips with the paraffin wax or silicone spray.

If it is the top sash that's loose, lower it and repeat the measuring and linoleum-strip installation process in the opposite order.

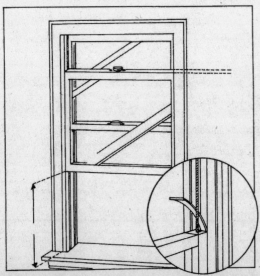

*A strip of linoleum in the window channel can tighten loose sashes.*

## Fixing Loose Frames

As a house ages, entire window frames work themselves loose because nails corrode and wood rots. To fix loose frames, you'll need the following tools:

- Hammer
- Caulking gun
- Screwdriver
- Framing square

You'll also need the following materials:

- Caulking compound
- Three-inch-long wooden shingles
- Oakum (hemp saturated with linseed oil—available in larger hardware stores or plumbing supply stores) or back-a-rod (a joint-filling plastic rope)

To fix loose frames, follow these steps:

1. Check the frame to see if it is square. If not, the window won't operate properly. To square up a frame, wedge wooden shingles in around the sides of the frame between it and the house frame (you may have to remove some molding to do this).
2. When the window is square, drive nails or screws through the window frame and the shingle wedges into the house frame.
3. Using the saw, cut off the excess shingle material as close to the frame as possible.
4. After the frame is solid, stuff oakum or back-a-rod into all spaces around the entire window. This will insulate the spaces and provide a support for the caulking material.
5. Caulk around the entire window frame.
6. Replace any molding you may have removed.

### Adding Weatherstripping

Weatherstripping is insulating material in thin strips, designed to block heat loss through narrow window-sash cracks and through the spaces around the edges of doors. You can get weatherstripping in several forms—spring-metal weatherstripping is the longest-lasting, but the most difficult to install. Adhesive-backed foam-rubber weatherstripping and adhesive-backed felt strips are easy to install, but they wear quickly. Rolled-vinyl weatherstripping is longer lasting than foam rubber or felt and is as easy to install.

### Hint

Check your weatherstripping every year. Weatherstripping wears out faster than most other insulation, and you'll need to replace it periodically to maintain your fuel savings.

Measure your windows and buy the appropriate amount of weatherstripping. Remember to install weatherstripping along the sashes and the sash-frame tops and bottoms. Carefully follow the manufacturer's instructions with whatever type of weatherstripping you buy to get the most effective insulation bang for your buck.

### Cost and Savings

**What Will It Cost?**

Weatherstripping costs about $15 a window if you do it yourself and about $100 a window if you have it done professionally.

**How Much Will I Save?**

A good, tight-fitting job of weatherstripping on all doors and windows will probably save you 10 percent of your home heating bill every year from now on!

## Chapter 3: Keeping Out the Cold 53

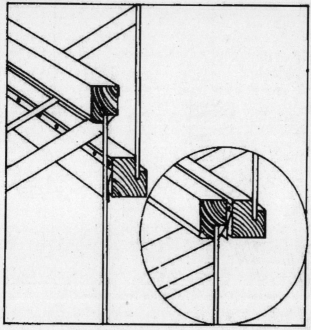

*Spring-metal weatherstripping on window rails.*

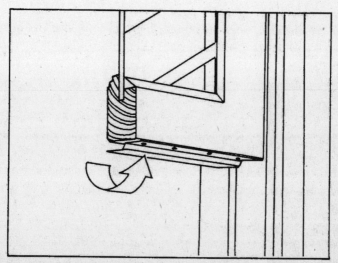

*Spring-metal weatherstripping on window-frame bottom.*

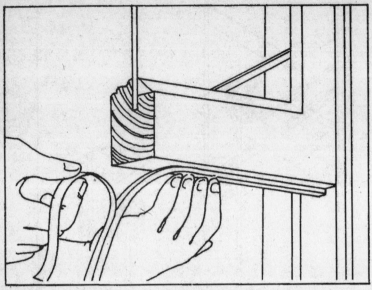

*Pressure-sensitive weatherstripping on window-frame bottom.*

## Applying Caulking

Half the fuel used in the average American residence for heating is wasted! That's right, 50 percent of the warm air the average furnace is making slips outside through cracks, leaks, and inadequately insulated roofs and walls. Plugging those leaks makes $ense!

Caulking seals your house's seams against heat loss. In addition to caulking around window frames, you should caulk wherever two different materials or parts of the house join. So, as long as you're going to buy caulking cartridges, you ought to check the area where the chimney and roof shingles meet, between roof dormers and shingles, all roof flashing, the undersides of eaves and gable moldings, between masonry steps and the main structure of the house, at corners formed by siding, and between siding panels and any protrusions from main structure of the house, such as hose connections, outside electrical panel boxes, and ventilators.

##  Hint

Caulking eventually wears out. Check caulked areas every year before the heating season begins and recaulk where needed. To extend the life of your caulking compound, consider a coat of paint. Painting your caulking compound will seal it against the wind. (Besides, it'll look better.)

How much caulking compound do you need? That depends on the job. You'll need about ...

- Half a cartridge per window.
- Half a cartridge per door.
- Two cartridges per chimney.
- Four cartridges for the foundation sill.

*Where you'll need to caulk your house.*

To do any caulking, you'll need the following tools:
- Chisel
- Medium flat screwdriver or putty knife
- Wire Brush
- Heating pad
- Pocketknife
- Caulking gun
- Old rags

You'll also need these materials:
- Cartridges of caulking compound
- Oakum or fiberglass
- Naphtha

To apply caulking, follow these steps:

1. Warm the caulking cartridges in a heating pad to make the compound easier to apply.
2. Using naphtha and a putty knife or screwdriver, clean the area you're going to caulk, removing old paint, dirt, and deteriorated caulk.

   **Note: Naphtha is flammable—don't smoke!**
3. Place the caulking cartridge in the gun and with the pocketknife, cut the tip off the cartridge at a 45-degree angle and insert the screwdriver to break the seal inside the cartridge.
4. Pull the trigger and squeeze out the compound, *pushing*, not pulling, the gun into the frame of the window. Drawing a good bead of caulk may take a little practice, but you'll get the hang of it. A good bead, incidentally, overlaps both sides of the crack for a tight seal.
5. For cracks that are too wide to be covered by the bead of caulking compound, fill in with oakum or fiberglass, and finish the job with caulk.

## Caulking Chart

| | Latex | Vegetable Oil | Silicone | Butyl Rubber | Nitride | Acrylic Polymeric | Polysulfide |
|---|---|---|---|---|---|---|---|
| Tack-Free Time | 15–35 Min | 2–26 Hrs | 60 Min | 30–90 Min | 10–25 Min | 10–30 Min | 24–72 Hrs |
| Ease of Use | Good | Fairly Good | Fairly Good | Fairly Good | Fairly Good | Poor | Fairly Good |
| Longevity | 10 Years | 5 Years | 20 Years | 20 Years | 20 Years | 20 Years | 20 Years |
| Minimum Application Temperature | 40°F | 60°F | 5°F | 40°F | 35°F | Warm to 120°F | 5°F |
| Adhesion to: | | | | | | | |
| Wood | Excellent | Fairly Good | Excellent | Excellent | Excellent If Unpainted | Very Good, If Unpainted | Excellent If Primed |
| Metal | Poor | Fairly Good | Excellent | Excellent, If Unpainted | Excellent, If Unpainted | Very Good | Excellent |
| Painted Surfaces | Excellent | Fairly Good | Excellent | Fair | Fair | Very Good | Do Not Use |
| Masonry | Good | Fairly Good | Very Good | Excellent | Excellent | Good | Excellent, If Primed |

>  **Caution**
>
> Lead-base caulk is toxic. It has been banned, but some stores still have stocks of it. Check cartridge labels and don't buy lead-base caulk.
>
> Also, if you have to use a ladder to caulk your windows and roof areas, be sure the ladder is level, and block it in place. Have someone hold it steady, if possible. When you finish one window, get down and move the ladder—don't stretch out for hard-to-reach spots. Make a sling for your caulking gun, so you can use both hands while climbing the ladder.

> **Cost and Savings**
>
> **What Will It Cost?**
>
> A caulking gun costs under $10. The cartridges can cost up to $6 each. Using the most expensive caulking material, it shouldn't cost you more than $120 to caulk all open cracks around your windows and other spots.
>
> **How Much Will I Save?**
>
> A well-caulked house will cost about 10 percent less to heat.

### Using Shades, Bands, and Drapes

Shades, blinds, drapes, and shutters can save you heating dollars, if you use the right types at the right times.

Earlier, we talked about keeping shades or drapes drawn on northern exposures during the winter, since you don't get much sunlight through them and they tend to conduct heat out of the house.

At night, when there's no sunlight to gather, you're best off shutting all your drapes, shutters, or blinds to create that extra layer of material over the windows and minimize heat loss.

(In summertime, you can maximize cooling and save on air conditioning by reversing the procedure—open the shades, windows, shutters, and so on at night to let in the cooler night air, and button them up in the daytime, to keep out the sun's radiant energy.)

*Make sure drapes cover the entire window.*

Made in the shade .... But what kinds of shade?

Naturally, the tighter the weave of your drapes, the better they'll work as insulation. Lined drapes are more effective than unlined drapes. Roll shades, if pulled all the way down, are more effective than venetian blinds.

And a good, tight fit is important, too. Make sure those shades and drapes leave virtually no space for warm air to escape at the sides, top, or bottom.

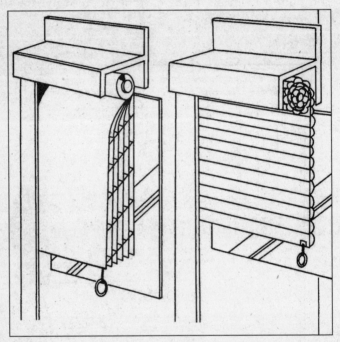

*Multiple-layer insulating window shade (left) and dead air space insulation window shade (right).*

There are two basic types of *insulated shades* on the market. Both are mounted in tracks along the sides of the window to eliminate air leaks at the edges. One type has a high R-value and is formed of multiple layers of plastic film. When drawn, the layers of film in the shades create multiple dead air spaces. A second type resembles venetian blinds when drawn, but the slats are actually rounded on the window side to create dead air spaces. Unlike blinds, however, the slats are connected by flexible joints. The dead air space in each slat, plus the dead air space between the slats and the window, give the shade its insulating properties. To let in light, the slats roll up into a valance above the window.

## Upgrade to Storm Windows and Doors

Now that your windows and doors are straight, tight, clean, caulked, and weatherstripped, it's time to consider a final step in equipping them to save you fuel dollars—storm windows and doors.

### Hint

**Check Those Old Storm Windows**

You already have storm windows? Good! You've had 'em for years? Good! Or is it? Check them. Old storm windows spring air leaks. Check to see that the screws holding them are still tight. And check to see if the aluminum has been buckled by the wind, creating gaps between glass and frame. Caulk gaps from the inside with a clear-silicone caulking cartridge. Also, check to see that the weep holes are clean and open. If you've got older storm windows without weep holes, you can make your own with your electric drill.

Storm windows and doors work by creating a pocket of dead air between the primary window or door and the outside. The dead air acts as insulation, preventing heat loss or cold air infiltration.

Dozens of manufacturers turn out a bewildering number of storm window models. It's easy to become confused and just buy by price. But the cheapest isn't necessarily the best bargain, and the most expensive doesn't necessarily assure you top quality. Storm windows and doors must be carefully chosen and carefully installed. An improperly designed or poorly fit storm window won't give you anywhere near the heat savings you're paying for.

When buying storm windows, check them for strength and appearance. The corners should be airtight and strong. They should have "weep holes"—tiny vents or drains that allow water condensation to escape from the dead air space. Without weep holes, the water will collect on your windowsills and rot them. Check the hardware—it must be as well-made and as sturdy as the rest of the window.

### Hint

Look for the Energy Star label on any storm window you buy. The label means the windows (and storm doors as well) exceed performance standards set by the Department of Energy and the Environmental Protection Agency.

Be certain that the storm windows you select have permanent weatherstripping or a vinyl gasket to seal the crack between them and the primary windows.

And remember, anodized- or baked-enamel-finish windows will look better longer. Bare aluminum windows will deteriorate faster and become quite unpleasant to handle as oxidation pits and mars them.

Shop around and shop carefully. Take your time. You're making a big investment and you don't want it to turn into a big mistake. Storm windows and doors can pay bigger dividends if you're a careful consumer. If you don't want to buy both storm windows and storm doors, bear this in mind: The windows will save you more than the doors. They fit tighter and can be left closed all winter long. Also, if you don't want to lay out the cash to do all your windows this year, you can do your windows over several seasons. If you want the most effect for the least expense, do your north-facing windows first. Next season cover the east- or west-facing windows. Leave the southerly exposures for last.

There are two basic kinds of storm windows available—and a third, temporary type you can make for yourself at very, very low cost. The basic types of "store-bought" storm windows are single-pane or single-sash windows and combination windows.

### Hint

Storm windows reduce the passive solar-heat-collecting ability of your windows because the rays have an extra layer of glass to penetrate. So, keep those storm windows clean to maximize that free heat from the sun.

### Using Single-Pane Storm Windows

These windows are a single pane of glass, usually in an aluminum frame, made to measure for your windows. They are fairly easy to install yourself, although you will have to work on a ladder if your house has more than one story, or if you've got high primary windows. Some models can be installed from the inside.

Some hardware and building supply stores sell kits for making your own single-pane windows if you've got the time and ability. Or, you can have them made for you. Either way, accurate measurements are critical. If you're doing the measuring yourself, remember this: Just because you've measured one window, don't assume that you've measured every other similar-looking window in your house. Windows vary in size. And, in fact, many vary in measurements from top to bottom, so measure all sides of every window.

### Hint

Save yourself a lot of confusion at installation time. Number each window on your house and list each window's measurements according to its number. Then number the storm windows when you get them. Installing them will be a matter of matching numbers, rather than a hit-or-miss proposition, trying to match various storm windows to various primary window frames until you get the right ones lined up.

Take your measurements to your building supply or hardware store, and be sure the salesperson copies them accurately when ordering your storm windows. Keep a copy of the measurements and compare your copy with the delivered storm windows to be sure they're right.

Mounting hardware varies from company to company, so follow the manufacturer's directions carefully if you're mounting your single-pane storm windows yourself.

Single-pane storm windows are less expensive than combination windows, but they have one major disadvantage: They must be removed every year at the end of the heating season. You get greater convenience—but at greater price—from combination windows.

### Using Combination Storm Windows

These windows combine two features—insulation for the winter and screens for the warmer months, so once they're installed, you won't have the hassle of taking them down every year. They remain permanently installed over your primary windows. When spring—or any other—cleaning time rolls around, you can easily open combination windows. You can also open them for ventilation at any time.

If you're really handy, you can install the combination windows yourself, but they are more difficult to put up than single-pane windows.

### Hint

> Be sure any combination storm windows you buy have permanent weatherstripping in the window channels as well as on the edges that will touch your primary windows.

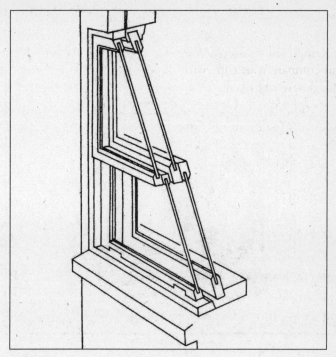

*Combination storm windows combine glass for winter and screens for summer.*

## Building Plastic Sheeting Storm Windows

If you don't want to spend $100 or so per window for storm windows, there is a temporary storm window you can make for yourself—with plastic sheeting.

You can easily install plastic sheeting storm windows and they're very inexpensive. They are not very attractive—some would call them downright ugly—but they are effective insulation and can be taken down anytime. (In fact, you'll have to take them down when the heating season ends.)

*Make-it-yourself plastic storm windows are easy and inexpensive, but effective.*

You can tape up the plastic sheeting inside your house over existing windows (so you don't have to work outdoors and can install them even on the coldest day). Or you can tack them up outside your primary windows.

To do the job you'll need the following tools:
- Scissors
- Ruler or yardstick
- Hammer (if you're tacking them to the outside) or double-faced adhesive tape (if you're taping them to the inside)

You'll also need the following materials:
- Six mil-thick polyethylene plastic (that's the thick stuff sold in rolls, not the food wrap or the board-like sheets)

- Tacks
- Thin wooden strips

If you are installing the plastic inside the house, do the following:

1. Measure each window.
2. Cut the plastic to size.
3. Cover the four sides of each window frame with the double-faced tape.
4. Remove the tape backing from the top only.
5. Line up the plastic and press in place at the top.
6. Remove the rest of the tape backing and smooth the plastic into place.

To install the plastic sheeting to the exterior of the house, do the following:

1. Measure each window.
2. Cut the plastic to size.
3. Tack plastic to the top of the window frame, using the wooden strips to hold the plastic in place.
4. Repeat the procedure on the sides and bottom.

## Cost and Savings

**What Will It Cost?**

Prices of storm windows vary widely, but we found combination storm windows for as little as $70 a window to just under $100 a window. For less than $75 you ought to be able to make your own temporary plastic storm windows for your entire house.

**How Much Will I Save?**

You should be able to cut your heating bill by 20 percent with well-fitting storm windows.

You can put temporary plastic storm windows over openings already protected by regular aluminum-and-glass storm windows. They'll fight heat loss even more. A good tip for northern exposures subject to strong winds.

### Using Storm Doors

When you're shopping for storm doors, inspect them the same way you checked out storm windows. Make sure they are of strong construction, have adequate weatherstripping, and that the frames are a treated aluminum. Storm doors, by the way, are also available with steel and wooden frames—and because doors get a lot more use than windows, you may want to consider these sturdier materials. Regardless of the material you select, look for the Energy Star symbol.

### Hint

Be sure that any storm door you buy has safety glass or clear plastic. That door will see a lot of opening and closing as well as rough treatment from the wind. You don't want the glass shattering in the middle of winter.

Unless you're extremely experienced and handy, you'll want a contractor to hang your storm doors. It's a much tougher job than installing storm windows. Before you let your contractor slip away, check the fit and the ease of opening and closing. It's much easier to get him to adjust it while he's still there than it is to get him to return to fix things.

### Buying Double-Glazed Windows

You can replace your old primary windows with newer double-glazed windows (many new homes in extreme-weather areas of the country are built with double-glazed windows). This is an expensive option, but will pay dividends for years to come, saving you heating and cooling dollars, and increasing the resale value of your home. As a bonus, you'll have a dramatic reduction in outside sound intruding into your home through double-glazed windows.

What is double-glazing? Simply put, it's two sheets of glass in the window casing with an airspace between, sort of a pre-made storm window effect on your primary window. Double-glazed windows are more effective than storm windows and they last longer because you never have to take them down or put them up, so they don't get the handling abuse suffered by storm windows.

Many double-glazed windows are on the market, so shop carefully and look for the Energy Star label to ensure that you'll get maximum performance from your new windows.

Most quality double-glazed windows are engineered so that they can be flipped inward into the room for safe and easy cleaning.

## Cost and Savings

**What Will It Cost?**

Double-glazed windows typically cost about $375 to $500 per window for wood, and $200 to $300 for aluminum or vinyl. If you're buying aluminum windows, make sure they have a "thermal break." That's a piece of insulating material between the outer and inner frame that prevents ice from forming around the interior window frame.

**How Much Will I Save?**

Retrofitting your whole house with double-glazed windows can save you as much as 25 percent of your heating costs and, in the summer months, you'll save as much as 15 percent on air conditioning as well.

## Weatherstrip Your Doors

Windows can be kept closed all winter, but doors can't—unless you're planning to hibernate.

Doors present the same heat-loss problems as windows, but, unless they're the rare, all-glass type, they don't offer the passive solar-heat collection opportunities of windows.

### Hint

It may sound silly, but the first "door" rule of heat conservation is, "Shut the door." A door swung closed and not latched is going to let a king's ransom in heat out into the cold winter air. So, shut the door—and teach your children to shut the door, too.

The best type of door for energy saving is a revolving door. But homes don't have revolving doors, so you've got to make do with what you do have.

If you've got a vestibule, use it as a cold-air lock. Close the vestibule door behind you before you open the front door. Also, turn off or remove the radiator or heat register in the vestibule. Use the area as a dead air space. Weatherstrip both the vestibule door and the front door, and insulate the interior walls of the vestibule.

### Hint

Protect yourself. If you're having windows or doors installed by a contractor, you should draw up a written contract, listing exactly what is to be done and how much it will cost. A reputable contractor will readily agree to such a document.

Almost anyone can weatherstrip a door—they're even easier to do than windows. But there are differences. The top and two sides of a door are frame stripped, while the door bottom requires special treatment, and you'll apply different types of weatherstripping using different techniques.

> **Hint**
>
> Keep your weatherstripping clean. It'll last longer and be more effective.

### Installing Pressure-Sensitive Foam Weatherstripping

This weatherstripping comes in rolls with adhesive on one side and an insulating foam on the other. It can be cut with a knife or scissors and installed simply by removing the protective backing from the pressure-sensitive adhesive and pressing it against the inside face of the door jamb. (Be sure to wash the jamb first with a good, strong detergent to remove all dirt and grease.)

Although this type of weatherstripping is extremely easy to install, it doesn't last very long—you'll probably have to replace it every two to three years.

*Pressure-sensitive weatherstripping is easy to install, but needs replacing every couple of years.*

**Hint**

You can reinforce the adhesive's sticking power by putting in a carpet tack or thumbtack every 10 inches or so along the length of the weatherstripping.

### Installing Spring-Metal Weatherstripping

This material is easy to install, but, unlike the pressure-sensitive weatherstripping, it lasts a long time. You'll need tin snips, hammer, nails, and a tape measure to do this job. Measure the door top and sides, cut strips to those lengths, and then tack the strips in place along door jamb. For a better seal, lift the outer edge of the strip with a screwdriver after tacking it in place.

*Spring-metal weatherstripping.*

## Installing Foam-Rubber Weatherstripping with Wooden Backing

This type of weatherstripping acts in the same way your refrigerator door gasket does. It comes in rigid strips and must be cut with a handsaw and installed with a hammer and nails. It is easy to install, and lasts almost as long as the spring-metal weatherstripping. As with the spring-metal weatherstripping, you cut it to length and nail it against the door jamb. You should hammer in the nails a foot apart or less. Here's a tip: If you put a bead of silicon glue on the door jamb before you nail the strips to it, the weatherstripping will last longer.

*Foam-rubber weatherstripping with wooden backing.*

## Installing Aluminum-Backed Vinyl Roll Weatherstripping

This material consists of a vinyl roll held rigid by an aluminum strip, which is nailed to the door jamb. It is easy to install and is long lasting. It's installed the same way as the spring-metal and foam-rubber wood-backed weatherstripping.

*Aluminum-backed vinyl weatherstripping.*

## Installing Sweeps

Sweeps are brush-like strips that are attached to the bottoms of doors to keep heat inside. They may be vinyl or nylon with either aluminum or plastic backing for rigidity and installation. They work only on flat thresholds and can become snagged on carpets or rugs. They are extremely easy to install, since the door can be left on during installation. Simply cut to proper length and screw on the bottom of the door. Check instructions of the sweep you buy to determine whether to mount it inside or outside.

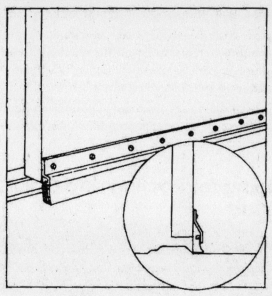

*A door bottom sweep.*

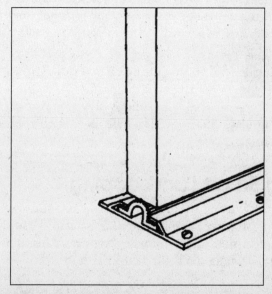

*A door bottom bulb threshold.*

### Using Shoes, Bulb Thresholds, and Interlocking Thresholds

Installing these devices requires you to remove the door and to trim the bottom to make room for the weatherstripping. While a do-it-yourselfer with good skills can install the first two, only a professional carpenter should attempt the interlocking threshold.

Finally, while you're weatherstripping all those doors and windows, remember the attic door. A lot of heat can slip through attic door cracks because hot air rises.

## Shopping for Roofing and Siding Materials

If you're about to re-roof or re-side your house, you're about to spend some really big bucks. Let's see how you can add a little energy-saving for your investment and make those bucks pay dividends by lowering your heating and/or cooling costs.

In the colder climates, choose dark roofing shingles. The dark shingles will absorb the sun's heat and act as solar collectors. In hotter, torrid zones—where cooling costs you big-time—go for lighter colors, which will reflect the sun's heat away from your roof.

In any area, if you're putting on new siding, it will pay you to look for metal or synthetic siding materials with insulation backing. And, when the old siding's off the house, it's a perfect time to insulate those exterior walls or add to the insulation already on them.

## Using Smart Landscaping

If you were starting from scratch, here's the ideal way to landscape your property for maximum fuel savings: First, determine where your winter winds normally originate. Most homeowners will find the winds come out of the north or west.

If that's the case for your home, on the north and west sides of your house, plant conifer (evergreen) trees. They will act as windscreens and aid your exterior wall insulation. With that cold north or west wind blunted, you'll feel warmer inside. (You know the feeling personally. When a cold wind is blowing and you step behind a shelter of some sort that blocks the wind, you immediately feel warmer.)

On all other sides, plant deciduous (leaf-bearing) trees. In winter, when the leaves are off, the sun's warming rays will strike your roof and generate heat. And in summer, the leaves will shade your roof, keeping you cooler and saving you money on air conditioning. Around the house, close to the foundation, plant hedges or other dense bushes.

In winter, if you're in a rural area, you may want to buy some hay bales and make a small mound of hay around the house at the foundation. The hay will keep the wind from the house.

If a portion of your home's foundation is exposed to the elements, you may want to consider regrading around your house to cover that portion.

You can also supplement your furnace with the sun's energy by attaching a greenhouse to your home. On sunny days, the greenhouse will collect the sun's heat, and a simple fan arrangement can pump that warmth into your home. On cloudy days and at night, close off the greenhouse with storm doors to cut heat loss. Depending on the size, greenhouses cost between $1,500 and $6,000. But they can reduce your home heating bill by as much as 25 percent!

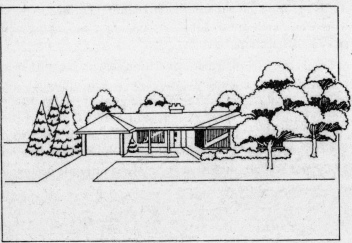

*Ideal landscaping protects the house from the wind but allows in sunlight in winter, and shades the roof in the summer.*

## HOW ABOUT A GREENHOUSE?

You can supplement your furnace with the sun's energy by installing a greenhouse attached to your house. On sunny days, the greenhouse will collect the sun's heat and a simple fan arrangement can pump that warmth into your home. On cloudy days and at night, close off the greenhouse with storm doors to cut heat loss. Depending on size, greenhouses cost between $1,500 and $6,000. But they can reduce your home heating bill by as much as 25 percent!

*A greenhouse can be a passive solar-heating unit.*

## For More Information

For further information check the following Internet resources:

| | |
|---|---|
| Energy Star | www.energystar.gov |
| American Architectural Manufacturers Association | www.aamanet.org |
| National Fenestration Rating Council | www.nfrc.org |
| National Wood Window and Door Association | www.nwwda.org |
| Andersen Windows | www.andersencorp.com |
| Pella Doors and Windows | www.pella.com |
| American Society of Landscape Architects | www.asla.org |
| U.S. Department of Energy, Energy Efficiency Clearinghouse | www.eren.doe.gov/erec/factsheets |

# Chapter 4

# Living Efficiently Year Round

Up to this point we've been concentrating on techniques for keeping heated (or, in summer, cooled) air in your house. But are you heating and cooling that air in the most efficient manner? In this chapter we'll give you some pointers for getting the most heating (and cooling) bang for your energy buck.

## How Do You Heat?

What kind of heating system do you have? If you answered gas, or oil, or electricity, or propane, then you only gave half an answer. What you named was your fuel. What you left out was your actual system. Basically, there are three types of heating systems: radiant, convection, and exchange. Some systems combine radiant and convection. Those aren't necessarily very sophisticated—cavepeople heated by a radiant system: They burned logs in the middle of their caves. (In addition to being unsophisticated, it was frequently dangerous because of the deadly gases given off by those burning logs. It makes you wonder how we as a species survived.)

That wood-burning fireplace of yours supplies radiant heat. The fuel—wood—burns, and what heat is reflected off the fire penetrates the room and warms you—if you sit close enough. Pretty, but not very efficient.

The ancient Romans went a step further. They used convection systems. They built big fires in their basements and detoured

*The Romans used a combination of convection and radiant heat in their ancient baths. The baths at Caracalla used technology like this.*

the flue pipes through the floors of their houses. Thus, the floors radiated heat to warm the rooms (maybe that's why men, as well as women, wore dress-like garments in those days—to let the heat rise up their legs).

After the Romans, in the dark ages, Europeans went back to radiant heat from fireplaces. In ancient castles you'll find walk-in-size fireplaces in almost every room. You can imagine whole forests going up in smoke while knights and their ladies stood practically *inside* those giant hearths, only to freeze a dozen paces from the fire.

Later on came steam heat. Wood or coal—and later oil and gas—were burned to heat water to its gasification point (steam is gasified water) and the steam was piped around a building and, through radiators and pipes, gave off heat. This was a combined radiant and convection system.

However, water has to be heated to more than *212°F* to turn it into steam. Another radiant and convection system is hot-water heat. Here the water is heated to only about *160°F* and piped around a house. It is quieter than steam and the heating system is simpler.

Forced-air heat, the most common form in this country, is generally a heat exchange and convection system. In the exchange, cool air is heated, and it circulates through a room. As it begins to cool, it is ducted back to the exchanger (the furnace) to pick up more heat.

Why do you need to know this? So you'll understand how your system works and understand why any obstruction that interferes with the exchange of heat will make the furnace work longer and harder and cost you more money.

Furnaces are generally sophisticated pieces of equipment. Heating systems are delicately balanced interconnections. You may be wary of tackling your system yourself. Well, you're right, for the most part. The big jobs aren't for home handypeople. But there are a few steps you can take to make your system work better. And the rest has to be done for you by a professional.

The first and easiest thing to do is to ...

## Turn Down That Thermostat

The thermostat is the brain of your heating system. It is actually a switch that turns the heating system on and off as it measures the temperature around it. It turns your furnace on when the temperature drops below a pre-set comfort level and it turns the furnace off when the comfort level is reached.

A heating thermostat does not work like the gas pedal on your car. Pushing it up to 85° does not make the heat come up any faster than if you set it at 68°. What you will do is overheat your house and waste your energy dollars. (And if you happen to have heating in your garage—turn it off! Your car's comfort isn't worth the price you're paying.)

If you're used to setting your thermostat at 72°, you can easily learn to live with a lower temperature of 70° or even 68°, especially if you wear a sweater indoors during the winter months.

At night, when you're under the covers, you can drop the temperature even further. If need be, you can plug in an electric blanket to keep you comfortable in a cold room. Turn down your thermostat and turn up your electric blanket. It costs less to heat a bed (even electrically) than to heat a whole house. Now, there's an old myth that says it doesn't pay to lower the thermostat at night because it takes extra energy to warm the building up to a comfortable daytime temperature. That's not true. Turning down that thermostat at night can save you anywhere from 10 to 15 percent on your heating costs.

> ### Cost and Savings
>
> **What Does It Cost?**
>
> Nothing.
>
> **How Much Will I Save?**
>
> Depending on your location in the country, you can save as much as 20 percent on your heating fuel bill by lowering your thermostat.

In general, for every degree you lower your thermostat in the winter you will save about 2 to 3 percent in heating costs! A drop of 4 degrees will save you 2 to 8 percent. A drop of 5 degrees, 10 to 15 percent.

You'll want a gadget that will lower your thermostat automatically at night and raise it back up just before you're going to get out of bed in the morning. Such a device is called a "setback." If work and school schedules mean no one is home during the day, you'll want to buy a "double setback" that will lower the temperature at night, raise it in the morning, lower it again after everyone has left the house for the day, raise it just before the family returns home, and dip it again at bedtime.

> ### Hint
>
> If you have central air conditioning, be sure the setback device you purchase can control that system as well so you'll get maximum economy during the cooling season.

For maximum flexibility, you'll want to look into the new electronic thermostats that enable you to program multiple setbacks over the week—so you can keep the heat on for more of the day on weekends, when everyone's home, and turn it down for the same hours during the week when no one's home. And there are even electronic thermostats connected to the Internet, so you can control them from your office computer or your laptop.

Depending on the level of sophistication, these devices can cost you anywhere from $55 to $145. Amazingly, the prices for today's most sophisticated electronic setbacks are no higher than the prices charged two decades ago for much less sophisticated thermostats.

### Caution

People can live in a house with lowered temperatures and get quite used to it. But it is important to remember that in some cases there are medical reasons not to turn that thermostat down.

Older people, particularly, are susceptible to hypothermia, a sharp drop in body temperature, which can be fatal.

People who have poor circulation or a history of hypothermia, and those taking certain families of prescription drugs, should take extreme caution not to let the temperature in their homes get too cold.

So if you fit the category or have someone living with you who does—don't lower that thermostat unless you first consult a physician.

If your thermostat is on an exterior wall or close to an entry door, it is needlessly costing you money. A thermostat should be on an interior wall, away from doors, windows, and heating units so it is not affected by drafts from open doors, or by air leaks through window cracks or faults in exterior wall insulation. If it's in the wrong place, have it moved. (It's not a job for a do-it-yourselfer.) Ideally, your thermostat should be in a large room, such as a living room or master bedroom, and should never be in a kitchen or bathroom.

## Use a Humidifier

All summer long, you complained, "It's not the heat that bothers me, it's the humidity." Now here we are, telling you that you need a humidifier in your home in the winter months. Why? Because moist or humid air holds heat better. And your body loses less heat through evaporation into more humid air than it loses into dry air. So, the humid condition that makes you less comfortable in the summer can help make you more comfortable in the winter. You can feel comfortable at lower temperatures if there's more humidity in the air—so a humidifier will help you turn down that thermostat and save fuel.

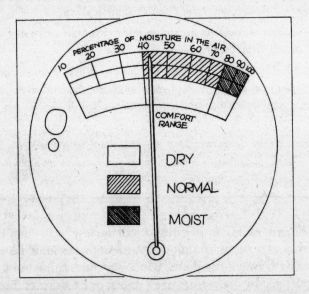

*A hygrometer.*

Before you buy a humidifier, go out and get yourself a hygrometer. This is a simple device that you hang on your wall near your thermostat. It tells you a room's relative humidity. For maximum comfort, you want the relative humidity to be between 45 and 75 percent. The "relative" in relative humidity refers to temperature. Cold air is capable of holding less moisture than warm air. A figure like 50 percent relative humidity means that the air has 50 percent of the humidity it can contain at that temperature. At 100 percent relative humidity, it's raining—a condition you want to avoid in your home unless you're a duck.

### Hint

**Keep It Clean**

A humidifier must be kept clean to operate efficiently. Lime and mineral deposits can build up from the evaporation of the water in the storage tank. Add lime-combating tablets to your humidifier, wash out the filters, and wash away excess lime deposits regularly, following the manufacturer's instructions.

Once you have your hygrometer, you can adjust a humidifier to the proper comfort levels.

### Cost and Savings

**What Will It Cost?**

A built-in humidifier can cost you as much as $350. Portable units begin at about $150 and go up, depending upon size, capacity, and controls.

**How Much Will I Save?**

Keeping the relative humidity at the optimum level should save you 10 to 15 percent of your home heating fuel bill because you'll feel warmer at lower temperatures, and so you'll keep the thermostat lower.

Some forced-air heating systems have built-in humidifiers. If the one that came with your house is operating properly, you don't need portable units. If you have a forced-air system without a built-in humidifier, it's possible to add one, but that is a job for a professional. Instead—or if you have a hot-water or steam system—you can buy portable units in most hardware, housewares, and department stores.

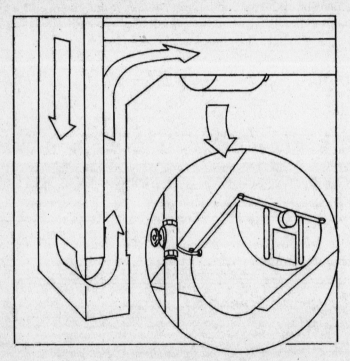

*Built-in humidifier in a forced-air system adds moisture to heated, dry air.*

## Make Your Old Furnace More Efficient

The following tips take a little more effort than turning down a thermostat but they're worth it.

## Remove Obstructions

All heating systems work best if the mechanism for delivering the heat is unobstructed. In forced-air systems, this means making sure that furniture, drapes, and rugs do not block the air registers. Placing a couch or bed over a register simply means you're heating upholstery, not a room. In steam and hot-water systems, remove enclosures from the radiators. It was once fashionable to build a nice, square box around radiators to hide them. Those boxes also absorb and obstruct a lot of heat that could be put to better use keeping you comfortable.

Dirt and dust are also obstructions. Keep your radiators clean, and vacuum your heat registers regularly.

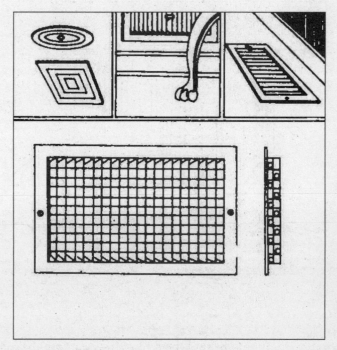

*Keep forced-air ducts clean and unblocked by furniture.*

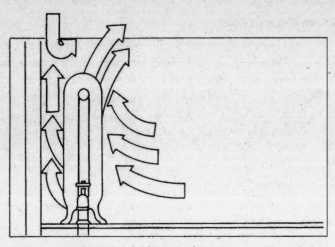

*Radiators warm by means of radiation and convection.*

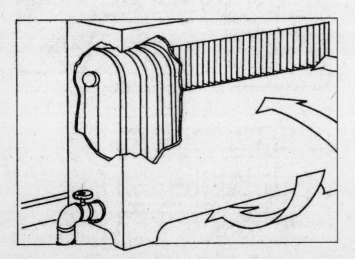

*Radiator enclosure's convection effect.*

You can make those radiators more efficient, too, by putting a sheet of aluminum behind them to reflect heat into the room. A more ambitious project can turn your old cast-iron radiators into modern, efficient heating units. You simply sandwich the radiator between two $1/8$-inch steel plates, bolted through the slots of the radiator. This converts the radiator into a convector so that the heated air circulates evenly and makes the room more comfortable.

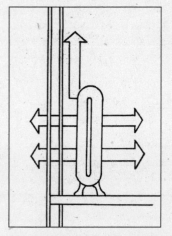

**HEAT LOSS THROUGH EXTERIOR WALLS (DRAWING A)**

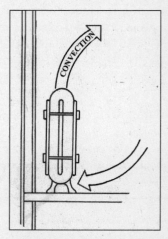

**HEAT LOSS THROUGH EXTERIOR WALLS (DRAWING B)**

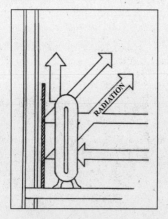

**HEAT LOSS THROUGH EXTERIOR WALLS (DRAWING B)**

*You can improve your radiators' heat output by placing a reflective aluminum sheet behind them or by bolting iron sheets on either side of them for maximum convection effect.*

### Lubricate Motors

If your heating system has motors, you'll want to lubricate them to increase their efficiency and life. Use nondetergent motor oil from your neighborhood hardware store and lubricate as indicated on the motor nameplates. This is an easy-to-forget chore, but one that will reduce friction and make your system work better. Some motors require no lubrication at all. If there are no instructions for lubricating on the motor itself, you probably have one of these.

### Hint

> Be careful not to over-lubricate your motors. Oil can drip into areas where it doesn't belong, reducing efficiency and even causing a burnout.

### Change Your Old Air Vents

If you've got a forced-air system, and some of the registers aren't adjustable, you probably find some rooms too cold and some too warm. Maybe you even things out by turning up the thermostat until the cold room is comfortable and open a window to cool down the hot room. Obviously, that's inefficient—you're heating the outdoors.

You can balance the hot and cold rooms more inexpensively by replacing the old registers with new ones that have adjustable vanes, and by using the vanes to turn down the flow of hot air into the warm room and turn up the flow of hot air into the cold room. Or, you can cover part of the register in the warm room with cardboard or masking tape, forcing more of the heated air into the colder room.

### Check and Repair Heating Ducts

The ducts bringing the warm air from your furnace to your registers probably run through your basement or through a crawlspace. Using a tissue held loosely in one hand, check these ducts for air leaks, especially at the joints. If you find any, repair them with duct tape available at any hardware store. In older houses, it's not unusual to find that ducts have separated at the joints, and heat that should be warming your family is filling the empty basement, crawl space, or attic.

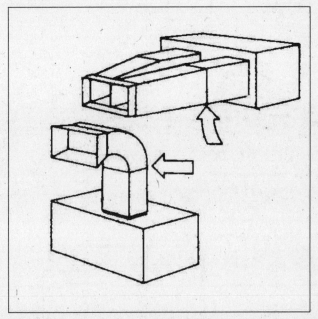

*Check hot-air ducts for leaks, especially at joints.*

## Change or Clean Clogged Filters

Just like your air conditioner, your forced-air heating system has a filter that prevents blockages by keeping out dirt and dust to prevent blockages. If that filter—located in the furnace or at a return register—gets clogged, the unit will operate inefficiently and will cost you more money. One recent study indicated that a severely clogged filter (a filter with only 20 percent capacity remaining) reduced heating efficiency by 25 percent! And it's so easy to avoid this problem that it's a joke.

Check your filter once a month, and clean or change it when it gets dirty. You can tell it's dirty by holding it up to a light. If little or no light shines through, it's dirty.

Change forced-air heating system air filters once a month during the heating season.

If you've got the kind of filter that can be cleaned, an inexpensive hair shampoo makes a good detergent for washing it. Don't wring the filter after washing; just squeeze it gently and let it dry naturally.

For less than $20, you can buy and install yourself a device that indicates when your furnace filter is clogged and needs changing or washing. This will save you the bother of having to open up the furnace front or the exchange register to check your filter visually.

Clogged filter indicator.

## Pitch Your Steam Radiators Properly

In a steam system, the radiators must be pitched slightly down toward the cutoff valve. This prevents the build-up of water in the radiator. If water comes in contact with the steam, it causes an annoying banging sound. This is known as a "steam hammer," and it's more than just bothersome, it's also an indication that the water is absorbing the temperature of the steam, so it costs more in fuel to heat your house.

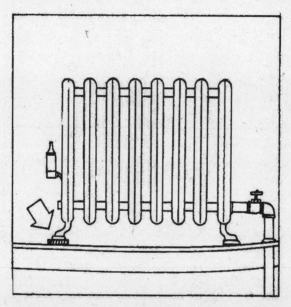

Use a small piece of wood to pitch your radiators and prevent a steam hammer.

If any of your radiators are not pitched properly, you can change that by wedging small pieces of wood or old bottle caps under the legs at the opposite end from the cutoff valve.

## Set the Cutoff Valve

The cutoff valve in a steam system is either on or off. Although there are many turns between the two, never leave the valve turned partway. That does not allow "a little heat" into the radiator; it causes banging in the pipes and it hampers the radiator's efficiency. "On" is counterclockwise—as far as it will go. "Off" is clockwise—as far as it will go.

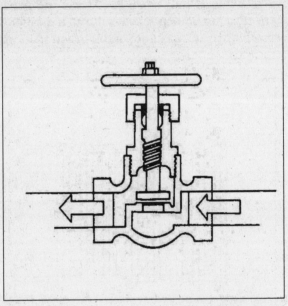

*Steam radiator valves are either "on" or "off." Turn clockwise all the way to allow the steam to flow, and turn counterclockwise all the way to shut the heat off. Check for leaks at the packing nut.*

### Hint

Check the cutoff-valve packing nut on each radiator. If any of them leak steam, the packing nut must be tightened. If it needs repacking, do it only with the system shut down, when there is no steam in the pipes.

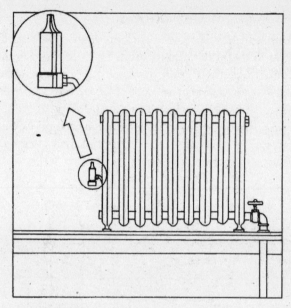

*Periodically replace air valves on steam radiators.*

## Check and Replace Air Valves

At one end of your steam radiator you'll find an air valve—usually it looks like a miniature factory whistle. Its job is to let the cold air out of the radiator as the steam comes in. It is engineered so that it closes when the hot steam comes in contact with it—to prevent steam from escaping.

When they're working properly, these valves allow the steam to fill the radiator more quickly, and under less pressure, saving energy. They must be checked and replaced periodically.

Valve replacement is a simple do-it-yourself project, because they screw off and on easily. Be sure to wrap the threads of the valves with dope tape to seal them. Do this job only with the system shut down, when there is no steam in the pipes.

All your air valves should be replaced at the same time, so you maintain a balanced system. And use only quality valves—they last longer, work more efficiently, and save you money in the long run.

## Bleed Hot-Water System Radiators

An air bubble can form inside your hot-water radiators and block the flow of heated water into the unit. To "bleed" the air out, find the air valve near the top on one side of your radiator. Usually you need a key or screwdriver (some bleeder valves have handles) and you'll also need a small cup to capture some of the dripping water. Open the valve slowly and let the air escape. When water begins to leak out, catch it in the cup and then close the valve. The radiator should work just fine.

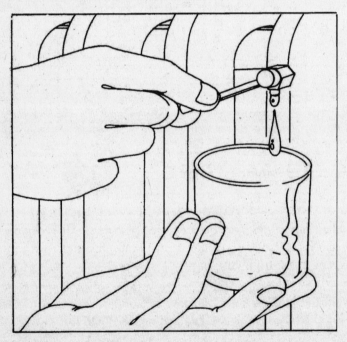

*Bleed air from hot-water radiators to prevent heat blockage.*

### Hint

**Paint It Black**

Radiators need painting? Consider flat black paint. Flat paint radiates heat more efficiently than satin or glossy paint.

## Have a Tune-Up

Just as your car runs more efficiently when its motor is in tune, your furnace, whether fueled by gas, oil, or propane, will also operate more efficiently when it's clean, lubricated, and adjusted for maximum performance.

In one survey, 97 percent of the oil burners checked weren't firing at the correct rate and thus were wasting energy. That's a staggering number—virtually every unit checked was wasting oil.

A simple system tune-up will pay for itself in no time. Savings from increasing the efficiency of your heating system could run as high as 30 percent!

*Don't try to do for yourself a professional job on your heating system. But beware of whom you do hire to do that work. In some jurisdictions, furnace technicians must be licensed.*

## Derating

A lot of oil burners and gas furnaces are too big for the size of the area they must heat. Thus, they operate inefficiently. They quickly build up enough heat to fog a house, and then they shut down.

To put it in automobile terms, it's like having a Cadillac motor in a Volkswagen. You won't get where you're going any faster, if you observe the speed limit, but you'll waste an awful lot of gas doing it.

Many furnaces with excess capacity operate only 30 percent of the time, even in the coldest winter months. During the off-cycle, heat is lost up the flue and through the furnace walls into the adjacent area.

One way of combating excess capacity is by "derating" your unit, putting a smaller nozzle on the burner, which means it will operate longer, but use less fuel, increasing efficiency and saving money.

## Get an Automatic Flue Damper

When your furnace is in its off-cycle—and even the most efficient units have off-cycles—heat is lost up the flue pipe. This can be remedied with a device called a flue damper. Simply put, the damper closes the flue when the furnace is off and reopens when the thermostat calls for heat. The reopening of the damper fires up the furnace. Thus, when no heat is being generated, the system remains closed. And when the fire restarts, the flue opens to exhaust the noxious and toxic gases produced by the furnace.

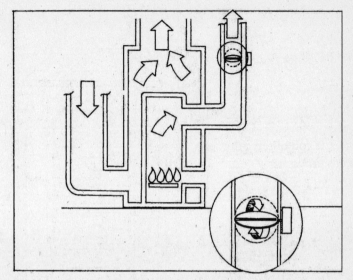

*How an automatic flue damper works. In off-cycle, damper closes flue, keeps heat in the furnace. In on-cycle, it opens to allow combustion gases to escape.*

## Have Hanging Baffles Installed

For old "pancake" boilers—like those in the following pictures—you can get more efficiency by having a hanging baffle installed.

Just as a hat keeps in your body heat on a cold winter's day, a hanging baffle keeps a "lid" on your furnace's fire—containing and concentrating the heat so that it isn't dissipated up the flue.

A hanging baffle is a stone box suspended by chains over the combustion chamber.

Modern furnaces do not need hanging baffles, but in an old unit a hanging baffle may save you as much as 10 percent in heating fuel. If you have an old furnace, have your serviceperson check it for a hanging baffle and, if it doesn't have one, consider ordering one or consider replacing it with a modern, new, fuel-efficient furnace.

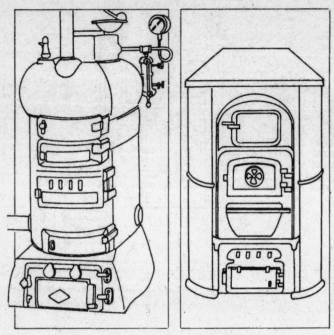

*Vertical or "pancake" boilers. Check for hanging baffles.*

## ⚡ Caution

An automatic flue damper should be installed only by a skilled, reputable, professional contractor. Dampers on gas-burning units should have an endorsement from the American Gas Association. Those on oil-burning units should have an Underwriters Laboratory label. *Do not* attempt to install your own flue damper and *do not* purchase any damper that does not have the necessary endorsement. An improperly designed or improperly installed flue damper can be very dangerous.

## Cost and Savings

**What Will It Cost?**

An automatic flue damper will cost you between $250 and $500 installed, depending on the model you choose and the difficulty of installation on your furnace.

**How Much Will I Save?**

Depending on your unit, you'll save 10 to 15 percent (units with longer off-cycles save more).

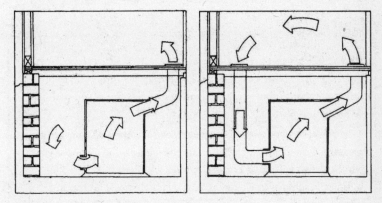

*Hot-air system without return air duct (left); hot-air system with a return air duct (right).*

### Install a Return Air Duct

On older forced-air and hot-air systems, the room air is returned to the furnace through a large vent in the floor. This air mingles with the cold basement air before it gets to the furnace to be reheated. On newer systems, the room air is returned to the furnace through a duct. Thus, the furnace doesn't have to work as hard since it's not warming a lot of basement air.

Also, as a fringe benefit, if you're no longer heating basement air, your house will be cleaner, as well as warmer. Without all that basement dirt and dust circulating through your forced-air system, you'll save your own personal energy by not having to work as hard cleaning. (And you'll save electricity, too, since you won't have to run your vacuum cleaner as much.)

If your system doesn't have a return air duct, consider having one installed. The savings should be substantial, although they are impossible to estimate, since they vary according to the conditions in each basement. In some extreme cases, however, you can cut your fuel bill nearly in half!

### Hint

**Be a Skeptic**

No matter how altruistic they may sound, bear in mind that your utility and your oil delivery firm are in business to *sell* you fuel, not to *save* you fuel.

When selecting someone to inspect, repair, or upgrade your heating system, you may be much better off engaging an independent who comes to you without a built-in conflict of interest. If his or her business is solely furnace maintenance and repair—and not fuel sales—he or she is earning money by saving yours.

That takes care of your primary heating system. It's time to see how you get some heat, as well as some light, from ...

### Your Fireplace

Remember those knights and ladies in that European castle, practically standing in their giant, walk-in fireplaces to keep warm? Well, the warmest part of the castle was really the flue—because that's where most of the heat went.

And it's little different today. Unless you live in the flue, you don't get a lot of heat from your fireplace. (And, of course, you couldn't live in the flue, even if you could fit, because in addition to heat, there's a lot of deadly gas coming up that chimney.)

That cheery fire may warm your heart on a cold winter night, but unless you take certain steps, that's all it'll warm. In fact, a fireplace—even with a fire going—can sometimes *cost* you heating dollars, because your gas- or oil-fired heat is escaping up that flue right along with the wood-fired heat. And after the wood fire goes out, that open flue is as bad as a hole in your roof.

An open flue can cost you more than $200 a year in fuel bills.

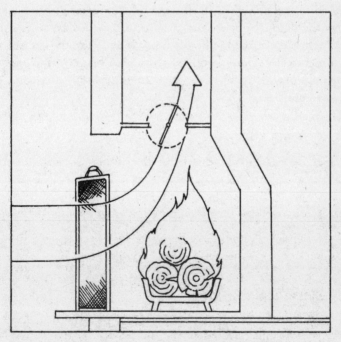

*Heated room air escapes upward with your open fireplace flue damper.*

### Close That Damper

A lot of fireplace flues have built-in dampers—a kind of door in the flue that closes it off to the outside. If your fireplace has such a door, close it when the fireplace isn't in use. But don't try to close the damper all the way when there is a fire going or even dying embers glowing—you'll fill your house with smoke and combustion gases.

If your fireplace was built with no damper, you can have one installed. The cost will vary between about $150 and $300, depending on the individual problems your contractor will encounter with your particular fireplace.

Even if you have a damper, however, there are drawbacks. If you go to bed at night with the fire still smoldering, you have to leave the damper open—and that means room heat will escape up the flue when the fire finally does go out. Also, it's pretty easy to forget to close most dampers—since they're up the flue, they're generally out of sight. There are some other remedies you may want to consider.

### Close Off the Fireplace

Shove fiberglass up the flue to block the loss of warm air. You obviously can't do this if the fireplace is to be used, but if you can do without a fire in the hearth, you can save heat this way. *Remember to remove the fiberglass before building a fire!*

Fiberglass is really a temporary solution. There are permanent solutions that are more costly, but are easier to live with.

### Buy Glass Fireplace Doors

Despite the claims of manufacturers, glass fireplace doors or screens don't appreciably increase the radiant heat your fireplace will give off. However, they will effectively block the exit of room heat up the flue, will enable you to go off to bed with a fire burning in the hearth, will increase safety because they lessen the risk of burning logs rolling out of the fireplace, and they don't diminish the beauty of a roaring fire. The least appealing aspect of a beautiful fire is that wood is a highly polluting fuel. In some smog-prone areas, limits are put on the use of wood-burning fireplaces and stoves. Check local restrictions before you start a fire in that hearth of yours.

*Glass fireplace doors prevent heat loss up the chimney.*

A fringe benefit is that glass fireplace doors cut the supply of oxygen to the fire somewhat, making the wood burn longer.

Glass fireplace screens cost around $200 and can be installed by most handy do-it-yourselfers. For others, installation charges are usually nominal, since it takes only an hour or so for a professional to install the screen.

### Use a Wood-Stove Fireplace Insert

From the outside, a wood-stove fireplace insert doesn't look much different from a fireplace outfitted with glass doors. But there is a major difference: A stove-like device has been inserted into the hearth and, through convection, as well as through radiation, it makes the wood fire far more efficient and heat producing.

These units, although costly, can eventually pay for themselves. Some offer fan attachments to enhance the convection function (sometimes called "induced convection").

## Wood-Burning Stoves

The wood-burning stove probably originated in the seventeenth century, but went into a decline in this country with the advent of coal, and later oil and gas heat.

However, wood is making a comeback, especially in modern, efficient wood-burning stoves. Typically, these units are used to supplement gas, oil, or electric heat in homes. They are far more economical in rural than in urban areas, because firewood prices are so much lower (if it isn't free), but even in bustling suburbs, more and more homeowners are installing wood-burning stoves.

*Wood-burning stove fireplace insert with a convector can heat a whole house.*

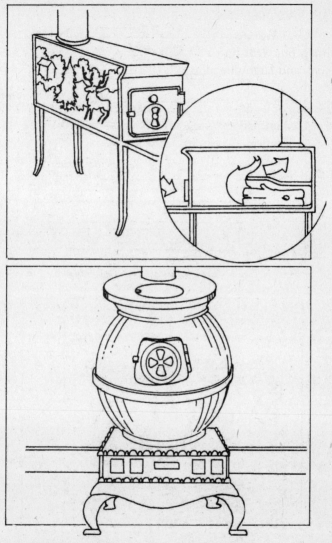

*Top: Horizontal wood-burning stove with bypass slows down the burning and puts out more heat.*

*Bottom: Round potbelly stoves can give off so much heat, with very little wood, that sometimes the room gets so hot you have to leave the house.*

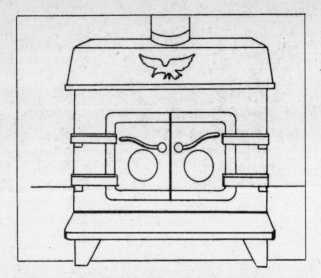

*This heavy-looking stove is designed to heat a whole house.*

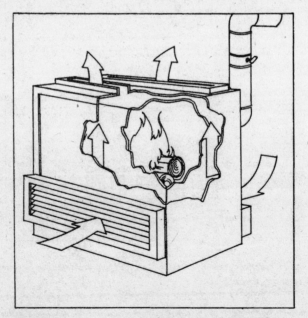

*How a modern convector and radiant-heat wood stove works.*

All wood-burning stoves supply radiant heat. The cast-iron or steel sides of the stove get hot and the heat radiates off the surface and into the room. As with any other radiant-heat method, the air is warmest closer to the source of the heat. In fact, stoves can become dangerously hot and should never be touched while a fire is burning in them.

Some of the newer stoves work by convection as well. These units have cavities around the firebox that create a convection flow of air. Cold air goes into the cavities at the bottom of the stove, is exposed to the heat of the fire (but not to the smoke or gases), and is expelled into the room at the top or sides. Some convection models have fans to induce more room air through the cavities.

### Wood-Stove Safety

Wood stoves can be dangerous. If you follow these basic safety tips, you should get a lot of secure enjoyment and supplementary heat from your wood stove.

- Don't use cardboard boxes or wooden boxes to store ashes. Use only metal containers.
- Don't use more than one heating unit for each chimney flue.
- Don't use masonry chimneys that do not have a flue tile lining.
- Don't use units where the flue pipe is loose or where it is not connected properly.
- Don't use chimney flues unless they are at least two or three feet above roof.
- Don't use chimneys if they do not have a fly-ash or spark screen.
- Don't place a stove on combustible floors such as wood, asphalt tile, carpeting, and so on.
- Don't use metal chimneys if they are not insulated away from wood-framed structures.

- Don't place furnishings or carpeting near through-floor chimneys.
- Don't place combustibles such as liquid, flammables, paper, cloth, and so on too close to your stove or chimneys.
- Don't dry wood, paper, cloth, or any combustibles on or near the stove.
- Don't use chimneys or stoves that are coated with creosote.
- Don't install any stove or chimney without consulting your local or municipal codes.
- Don't use liquid flammables to fuel or ignite stove.
- Don't use charcoal or coal—hard or soft—as a fuel in a wood stove.
- Don't use units unless they've been checked for cracks.
- Don't use stoves placed less than three feet from any combustible wall.
- Don't touch the surface of your stove while it's in use.
- Don't service hot stoves while wearing loose clothing.
- Don't permit children to play near unattended stoves.
- Don't use green wood.
- Don't use your wood stoves as an incinerator.
- Don't use stoves without having fire extinguishers or a bucket of sand or water nearby.
- Don't buy stoves that are oversized for the area to be heated.
- Don't use or install smoke pipes that extend more than 10 feet horizontally.
- Don't place stove pipes closer than 18 inches to any combustible material.
- Don't use plastic piping or any other combustible material for stove pipes.
- Don't use any stove unless it's been approved by a testing laboratory.

# Chapter 4: Living Efficiently Year Round

> **Hint**
>
> **Nuts to Your Fire**
>
> Does your fire dwindle down to virtually nothing after a few minutes? Is it lazy, lethargic, and not very bright? Well, nuts to it! Or more accurately, nutshells. Nutshells are hard and dense and burn brightly and hotly. Save nutshells and use them for both starting your fireplace fire and for reviving it when it begins to die out.

*Nutshells are hard and dense and burn for a long time ... as close to coal as you can get; but never use coal or charcoal, because their combustion gases are toxic.*

## How Economical Is It?

The cost of wood as a fuel varies so greatly around the country that it is impossible to generalize about how economical it is.

If you live in the heart of a big city where nothing grows but light poles, fire hydrants, and parking meters—and the nearest forest is some distance away—wood isn't going to be economical.

If you live in the backwoods, with a forest at your door, all the fuel will cost you will be the sweat of your brow (or maybe a little gasoline, if you use a chainsaw).

Wherever you live and however you get your wood, remember that dry wood is best, and that split wood burns better than unsplit, round logs. Dry—or seasoned—wood has generally been cut and stacked and left standing in a dry place for 10 months to a year. The dry wood will leave less soot in your chimney and will burn hotter.

Softer woods—pines, firs, other evergreens—start fast but burn very quickly. Hard woods—oak, ash, maple—are harder to start, but since the wood is denser, their fires last longer.

A cord of wood has about the same heat-producing power as 200 gallons of heating oil, when the wood is burned in an efficient stove. But what is a cord of wood? A cord of wood is 128 cubic feet of wood. Put in more understandable terms, it is a stack of logs four feet deep, four feet high, and eight feet long.

A good supply of dry firewood, an efficient wood-burning stove, and a few other precautions, can save your life—and your family's lives—in a fuel emergency. While it may seem unlikely that all sources of heating fuel will be totally cut off, it once seemed unlikely that we'd be paying more than two dollars for a gallon of gasoline. The unexpected has a nasty habit of happening—and, just in case, we've collected a series of live-saving tips in Appendix B, "Prepare for the Worst."

# Chapter 4: Living Efficiently Year Round

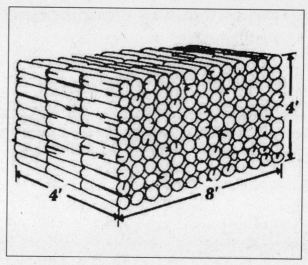

*A cord of firewood measures four feet deep by eight feet long and four feet high.*

## Characteristics of Commonly Burned Fuel Woods

| Species | Length of Burning | Ease of Starting |
| --- | --- | --- |
| **Hardwoods** | | |
| Oak | Excellent | Poor |
| Birch (white) | Good | Good |
| Ash | Good | Fair |
| Maple (sugar) | Good | Poor |
| Elm | Fair | Fair |
| **Softwoods** | | |
| Fir | Fair | Good |
| Hemlock | Fair | Good |
| Spruce | Fair | Good |
| Cedar | Poor | Excellent |
| Pine (white) | Poor | Excellent |

## Cool, Man, Cool: Air Conditioning and Ventilation

In the days of cheap energy, we grew to think of air conditioning as a necessity. But there are some of us who remember the days when it wasn't even a luxury—when it didn't exist at all. Somehow, humankind survived! So, in this day and age, we can certainly take a little more heat before we get out of the kitchen (or turn on our air conditioners); we can operate air conditioning at lower temperature settings; we can ventilate our homes naturally; and we can save a lot of money, too.

Even the most efficient air conditioners are relatively expensive appliances to operate since they are heat-transfer machines with heavy-duty motors. So the less we use them, the more we save. And the more efficient we make our air conditioners, the more we save.

### Tolerate Higher Temperatures

In the face of recent electrical energy shortages in California, the federal government raised the temperature in its buildings in the state to 78°F. That's warm, but livable. For every degree you raise the temperature in an air-conditioned room, you'll save 2 to 3 percent of your cooling costs. So, if you go from the old standard of 72° to the new standard of 78°, you'll save a whopping 12 to 18 percent!

**Note:** An air-conditioning thermostat does not work like the gas pedal on your car. (I think we spoke about this in the heating department.) Pushing it down to 60° does not make the AC system cool you any faster than if you set it at 73°. What you will do is over-cool your house and waste your energy dollars.

### Service Your Air Conditioner

If you've got an air conditioner, you're going to use it. So let's use it in the most economical way possible. Keep it running smoothly, cleanly, and efficiently, and it'll do the job of cooling you a lot more cheaply than if you don't maintain it.

And *you* can do a lot of the maintenance yourself, whether you've got central air conditioning or individual window units.

All air conditioners should be serviced at least once a year. Motors should be lubricated, if required, coils cleaned, and filters checked and changed.

### Coils? Motors? Filters? What and Where?

Well, all air conditioners operate as heat extractors and humidity extractors. They suck warm, moist air out of the room or house, remove the heat and moisture, and dissipate them outside the structure. Dirt will hinder the flow of the hot air—making the unit work longer—and will prevent condensation of the moisture. In air-conditioning terms, cleanliness is not only next to Godliness, it's also *cool!*

> **Caution**
>
> Be sure you've turned off and unplugged the unit when you remove and replace the filter. A running, open air conditioner has hazardous fan blades spinning so fast you may not even see the motion. You can really hurt yourself if you don't shut the unit down before you begin working on it.

- ◆ **Filters.** Filters in window units should be cleaned or replaced once a month during the cooling season. Foam filters can be cleaned by washing them in liquid detergent or—even better—an inexpensive shampoo. You can replace the washed filter in the unit while it's still damp. (But remember to shut off the unit before you extract the dirty filter and to leave it off until you've put the newly cleaned filter back in.)

Replace the filters on central air conditioners once a month. Many units have two filters, and replacements should cost you less than $2 for each filter. Watch for the arrow mark on the edge of the replacement filter. The arrow should point in the direction of the air flow. Again, be certain the unit is shut off before you open it up to extract the filter. Anyone can do this job—it's simple. But it's also easy to forget. Remember those filters!

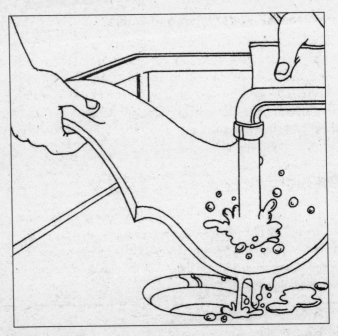

*Wash air-conditioner filters once a month.*

◆ **Coils.** The air conditioner coils are heat exchangers. Interior coils absorb heat, and exterior coils expel it. Dirt build-up on the coils acts as insulation, hindering both functions. Clean internal coils with a damp rag or a soft brush, and then vacuum. And you can hose down the external coils while you're watering your lawn or washing your car.

*Keep your air conditioners clean.*

## Economy Runs

On really hot days with low humidity, run the air conditioner with the fan set on high. In very humid hot weather, set the fan at low speed to provide more moisture removal, but less cooling. You can feel comfortable at higher temperatures if the air's less humid.

## Shut It Off

It is not more economical to leave an air conditioner on all day. A machine doesn't have to work longer to cool a hot room than to maintain a constant cool temperature in a room. So if you're leaving the house—or temporarily quitting an air-conditioned room—save yourself some dollars by turning off the air conditioner.

## Be a Cool Customer

When you go shopping for an air conditioner, check the EER (Energy Efficiency Rating) of the unit. The EER is designated on a bold black and yellow label affixed to every air conditioner sold in the United States. The higher the EER, the more efficient (and money-saving) the unit. Top-performing air conditioners have earned an Energy Star label by meeting high economy standards set by federal agencies.

Buy air conditioners that are adequate for the job, but not overpowered machines. The smallest capacity unit for the job is the best one because it runs continuously, always removing humidity. An overpowered unit for the space will shut down when the target temperature is reached, and the humidity will climb quickly, making you feel uncomfortable. More cooling power than you need is *inefficient* and wastes your cooling bucks, both when you purchase your unit and when you run it.

### Hint

**Storm Windows in Summer?**

When running your air conditioners, be sure to keep doors and windows closed. And, ironically, those storm windows you installed to keep out the stiff winter winds can keep the cool air-conditioned breezes *inside* during the summer, so think about leaving them up year round.

## Shade It

The ideal location for an air conditioner is the north side of your building or in the shade. If that is impossible, consider installing an awning over an air conditioner. The less heat hitting the unit outside the house, the easier it is for the machine to dissipate heat from inside the house.

*Shading an air conditioner will increase its efficiency.*

## Be a Fan of a Fan

Remember those lazy-moving ceiling fans in *Casablanca* with Humphrey Bogart, Ingrid Bergman, Peter Lorre, and Sidney Greenstreet? They were there mostly for atmosphere. But if *you* install one or two, they'll stir up a pleasant breeze to combat the summer's heat. They really work. (Would Bogie install a dud in Rick's Café Américain?) You can run a ceiling fan at the same time that you run your air conditioner. In fact, ceiling fans *increase* the comfort level, enabling you to raise the air conditioner's thermostat temperature.

 **Hint**

### Fringe Benefit

In winter, a ceiling fan running at it's slowest speed will force heated air—which naturally is forced to the ceiling—down to you from the ceiling, helping you keep that thermostat down.

Alternatively, you can also install an attic fan. You do remember attic fans, don't you? They used to work. They still do. They work even better today, because they've been scientifically improved. And an attic fan can ventilate your whole house so that on those marginal days, you won't need the air conditioner at all; you'll be able to cool by ventilating. A big attic fan has a motor rated at about $1/4$ to $1/3$ horsepower. By contrast, an air-conditioner motor can be anywhere from one full horsepower to three or four horses. It's obvious that it's less expensive to feed $1/3$ of a horse than three horses, so a fan can save you a lot of money (an air conditioner can use as much as 10 times the expensive electricity that a big attic fan uses).

### Hint

**One Cool Device at a Time, Please**

It might seem that by running both your attic fan and your air conditioner at the same time, you'll get double the bang for your buck. *Wrong!* The fan will merely suck the conditioned air right out of the house and you'll be paying to cool the great outdoors. If you're using one, turn off the other.

For a really good do-it-yourselfer, installing an attic fan is a challenging project. For most homeowners, though, it's a job for professionals.

### Caution

Your attic fan should have a thermostat—one that switches itself on when the temperature in the attic climbs above 100°F. But it *must* also have a safety disconnect device so that it shuts off in the event of a fire in the house. A running ventilator fan will make a fire more intense by drawing oxygen into the building.

## Chapter 4: Living Efficiently Year Round 123

Where do you put that fan? Well, the most effective spot is on an exterior wall, either at the gable end or between the rafters of your roof. Units installed in the attic floor aren't as effective, and they spread attic dirt into the living area of the house.

*Ceiling fans help your air conditioner keep you comfortable.*

## Explore Additional Ventilation Alternatives

You'll have to take pains to see that all closed spaces and dead air spaces in your home—such as attic areas, crawlspaces, cocklofts, and underground vaults—have some ventilation all year round to permit a free flow of fresh air and to allow moisture to dissipate. The ventilator can be as simple as a small grill covered with screening to prevent insects and vermin from getting in your home.

## Nature's Own

Hot air rises, right? Well, not exactly. Hot air is forced up by cooler air—which is heavier.

Why are we telling you all this? Why do you need to know it? So you'll be able to follow the reasoning behind a natural ventilation system that uses no fans at all.

On the north side of your home, on the lowest floor, open all the windows. Now, on the top level, south side, open the windows. The cooler, north-side air will enter the house and force the warmer room air up and out the south-facing window.

You can enhance that effect by draping wet sheets over the outside of the open north-facing windows and keeping them wet. This is especially effective in low-humidity areas.

*Natural ventilation flow.*

## Chapter 4: Living Efficiently Year Round

**Wash Your House**

Another trick is to hose down the exterior of your house on particularly hot days. As the water evaporates, it draws heat out of the building.

Those are some ideas to keep you cool.

## For More Information

For further information check the following Internet resources:

| | |
|---|---|
| Energy Star | www.energystar.gov |
| Department of Energy, Office of Building Technology | www.eren.doe.gov/buildings |
| Honeywell, Inc. | www.honeywell.com |
| Gas Appliance Manufacturers Association | www.gamanet.org |
| Heatilator Corp. (fireplace inserts) | www.heatilator.com |
| Slant/Fin | www.SlantFin.com |
| Air Conditioning and Refrigeration Institute | www.ari.org |
| Carrier Corporation | www.carrier.com |
| Fedders (air conditioners and fans) | www.fedders.com |
| Trane | www.trane.com |
| Hunter Fan Company | www.hunterfan.com |
| Casablanca Fan Company | www.casablancafanco.com |

# Chapter 5

# Stretching Your Energy Dollar—at Home and on the Road

It seems we have more and more energy-consuming gadgets around our homes—and, on the road, a distressing fact: Cars are getting bigger, which means thirstier. You can get a graphic idea of just how many more electricity-consuming appliances you have today than just a few years ago by counting the number of multiple-outlet power strips you've been forced to place around your house to supplement your normal wall outlets. Years ago the rule was to have a single outlet on every wall; then people needed double outlets. Now, the new rule is to have multiple outlets every 10 to 12 feet.

There are ways to save energy and not give up your twenty-first-century lifestyle. In this chapter we give you a few energy-saving (and money-saving) ideas.

## Saving Around the House

In any family there can be children who are picky eaters, hardly touching their food, and other children who seem to have bottomless pits for stomachs.

It's the same way with those dozens of electric appliances you've got around your house. Some of them have really voracious appetites, while some use very little power.

One way to save is to use the energy-gorgers as little as possible. The following list includes the most common appliances along with their annual estimated consumption in kilowatt-hours.

To work out the actual cost to you, check your electric bill to see what your cost per kilowatt-hour (kWh) is. Now multiply the rate by the kilowatt-hour consumption of the appliance, and you'll have a ballpark estimate of what your actual dollar costs are for each appliance.

Of course, the cost will vary according to the size of the appliance and the amount of usage. If you leave your television on all day, you'll use more than the "average usage" figure cited in the following table. And if you have a huge refrigerator, you'll use more than the average refrigerator.

## Average Appliance Usage

| Types of Appliances | Estimated Annual kWh Usage |
|---|---|
| **Major Appliances** | kWh = 1,000 Watt Hours |
| Air conditioner (room)* | |
|   10,000 BTU | 2,520 |
|   7,000 BTU | 1,680 |
|   5,000 BTU | 1,260 |
| Clothes dryer | 993 |
| Dishwasher | 288 |
| Freezer (16 cu. ft.) | 1,190 |
|   frostless (16.5 cu. ft.) | 1,920 |
| Range with oven | 700 |
|   with self-cleaning oven | 730 |
| Refrigerator (12 cu. ft.) | 728 |
|   frostless (12 cu. ft.) | 1,217 |
|   w. freezer (12.5 cu. ft.) | 1,500 |
|   w. freezer—frostless (17.5 cu. ft.) | 2,250 |
| Washing machine | |
|   washing machine only | 120 |
|   incl. energy used to heat water | 2,500 |
| Water heater | 4,811 |

| Types of Appliances | Estimated Annual kWh Usage |
|---|---|
| **Small Kitchen Appliances** | kWh = 1,000 Watt Hours |
| Blender | 15 |
| Bread maker | 36 |
| Broiler | 144 |
| Coffee maker | 140 |
| Deep fryer | 83 |
| Egg cooker | 14 |
| Frying pan | 192 |
| Hot plate | 90 |
| Mixer | 13 |
| Microwave oven | 190 |
| Roaster | 205 |
| Sandwich grill | 48 |
| Toaster | 39 |
| Toaster/oven | 120 |
| Trash compactor | 50 |
| Waffle iron | 48 |
| Waste disposer | 30 |
| **Heating and Cooling** | |
| Air cleaner | 216 |
| Electric blanket | 147 |
| Dehumidifier | 3,240 |
| Fan | |
| attic | 360 |
| circulating | 43 |
| roll-away | 138 |
| window | 170 |
| Heater (portable) | |
| 500 watts | 360 |
| 750 watts | 495 |
| 1,000 watts | 720 |
| 1,500 watts | 1,080 |
| Heating pad | 10 |
| Humidifier | 163 |

*continues*

## Average Appliance Usage (continued)

| Types of Appliances | Estimated Annual kWh Usage |
|---|---|
| **Laundry and Cleaning** | |
| Iron | 60 |
| Vacuum cleaner | 48 |
| **Health and Beauty** | |
| Electric toothbrush | .5 |
| Germicidal lamp | 141 |
| Hair dryer | 180 |
| Heat lamp (infrared) | 60 |
| Shaver | 1.8 |
| Sun lamp | 16 |
| Treadmill | 228 |
| **Home Entertainment/Office** | |
| Answering machine | 84 |
| CD player | 120 |
| Computer game console | 144 |
| Fax | 84 |
| Personal computer | 144 |
| Television | |
|    Black and white | 120 |
|    Color | 440 |
|    VCR | 24 |
| **Housewares** | |
| Clock | 24 |
| Sewing machine | 12 |
| Vacuum cleaner | 48 |
| Central system | 96 |

*Based on 1,000 hours of operation per year. This figure will vary widely depending on geographic area and specific size of unit.

## The Codes of the West, East, North, and South

You can save power by buying only the most miserly appliances and then by using them correctly. And you don't have to spend hours and hours doing research on which appliances are the least expensive to feed; the government's done all the work for you—all you have to do is *read the label* and become a Watt Wizard.

- **EER (Energy Efficiency Rating)** is the code of the west and the east, the north and the south. It is a measure of energy efficiency for room air conditioners. Only models between 8,000 and 13,000 BTUs use this scale. (A BTU, or British Thermal Unit, is the universal measurement for energy-consumption. A single BTU is the equivalent of raising the temperature of one pound of water one Fahrenheit degree or 252 calories.) Units with an EER of 10 or greater are the most efficient.

- **SEER (Seasonal Energy Efficiency Rating)** is a measure of energy efficiency for central air conditioners. Units with a SEER of 13 to 16 are money savers.

- **HSPF (Heating Seasonal Performance Factor)** is a measure of energy efficiency for heat pumps when heating. A unit may have both an HSPF and a SEER rating. Heat pumps with an HSPF of 8 to 9 or greater will heat with less fuel. The same unit with a SEER of 13 to 16 is a better money saver.

- **AFUE (Annual Fuel Utilization Efficiency)** is a measure of energy efficiency for furnaces and boilers. (Only furnaces fueled by natural gas are used in this scale.) Furnaces with an AFUE of 85 and above burn less gas fuel.

- **Therms/Year** on gas-fired water heaters is a measure of energy use. Your utility company uses it to compute your bill. Water heaters with a consumption rating of 245 Therms/Year or lower are saving you money.

- **KWh/Year (kilowatt-hours per year)** on refrigerators, freezers, dishwashers, and clothes washers is a measure of energy (electricity) use. Units with less than 900 kWh/Year use less electricity.

*Very confusing*—we really don't understand why they can't make the scales uniform and a lot less complicated but that's the way it is.

So you must check out the "EnergyGuide" labels on all new appliances before buying; that is, if you want to save energy and money.

Under the law, manufacturers of major appliances must list an energy-efficiency rating on their products, so that you can compare them with similar products of other manufacturers and buy those that will be most economical to operate.

The Federal Government assigns an EER, SEER, AFUE, Therms/Year, and kWh/Year to each appliance, water heater, air conditioner, furnace, and heat pump, after running tests on it.

The EER is incorporated into the EnergyGuide (cq) label, a bold, black-on-yellow label, which *must* be displayed on these appliances:

- Central and room air conditioners
- Clothes washers and dryers
- Dishwashers
- Freezers and refrigerators
- Furnaces and other home heating equipment
- Humidifiers and dehumidifiers
- Kitchen ranges and ovens
- Water heaters
- Television sets

# Chapter 5: Stretching Your Energy Dollar

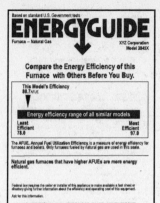
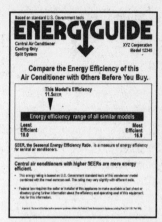
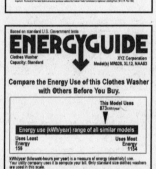
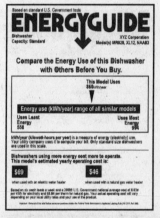
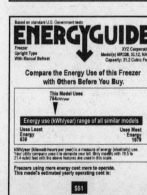
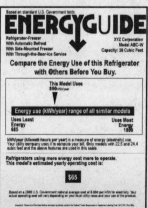

## 134  Save Energy, Save Money

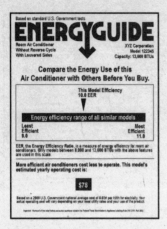
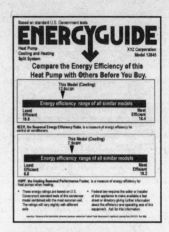
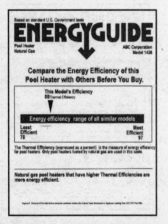
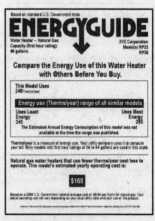

Look for the EnergyGuide labels on most major appliances. They can save lots of money if you buy the unit that uses fewer watts and can still do the job.

Let's take a closer look at a typical EnergyGuide label for a major appliance.

## Chapter 5: Stretching Your Energy Dollar

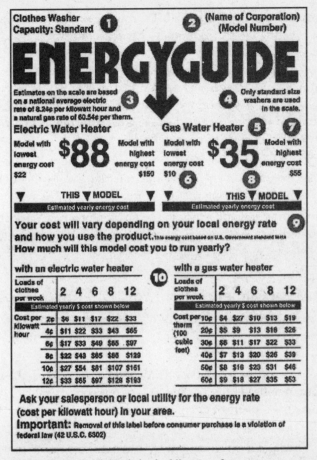

The EnergyGuide label includes the following information:

1. Type of appliance and capacity.
2. Name of the manufacturer and model number of this appliance.
3. National average cost of energy.
4. The models compared to develop the label.
5. Estimated annual cost of this model based upon national average energy cost.
6. Lowest annual operating cost for similar models.
7. Highest annual operating cost for similar models.
8. Marker arrows to show how the operating cost of this model compares to all similar models.
9. Scale that shows the lowest to highest annual operating costs.
10. Chart that shows the annual operating costs of this model based on different energy costs.

The easiest way to read the label is to look at the horizontal bar just above the label's midpoint. That tells you where the appliance you're considering stands vis-à-vis the least and most energy-efficient similar models. The label also gives an estimate of annual kWh usage, and the annual cost at several levels of kWh rates; so, if you bring your utility's kWh rate to the store, you can estimate your annual usage. Check the model's estimated yearly operating cost (box or indication).

Don't let the salesperson talk you into buying a cheaper appliance with an energy-voracious appetite—you will pay for it many times over during the life of the unit.

Of course, if you're not in the market for a new refrigerator, air conditioner, or range right now, you'll want to know about how to …

## Get More Efficiency from Your Old Appliances

We use energy to run a variety of appliances—everything from the essentials, such as water heaters, to the frivolous, such as the video and computer games.

The more we conserve on any of these appliances, the more we save in dollar terms and in terms of our country's dependence on foreign suppliers of energy fuels and our own supplies of fossil fuels.

The largest electrical energy eater among your household appliances is your water heater.

It probably doesn't cross your mind when you open that hot-water tap to do the dishes or to fill your bathtub, but 15 percent of all the electrical energy used in your home goes for making that water hot. (Many homes have gas water heaters, which use insignificant amounts of electricity to operate ignition systems. If you have a gas water heater, these tips apply, but—obviously—will have a positive impact on your gas bill, not on your electric bill.)

## Chapter 5: Stretching Your Energy Dollar   137

Here's how to save some money on your hot-water fuel bill:

◆ **Lower the water temperature.** You burn more fuel or use more electricity to get that water to a higher temperature. A setting of 160°F, while common, is too hot. You don't need water that hot, and you can scald yourself with it. So, for safety's sake, as well as for conservation's sake, you should lower that water temperature. How far? Well, dropping the temperature from 160°F to 140°F will save you about 18 percent of your water heating bill. Dropping it another 20 degrees, to 120°F, can save you another 18 percent. And 120°F ought to be adequate for most household needs, except for dishwashers, which generally need water of 140°F to operate properly.

 **Caution**

Think 120°F is not hot enough? Try taking a 110°F shower—you'll be able to stand that temperature for about two seconds before turning red and starting to scream in pain.

All water heaters have a temperature control on them somewhere. Some are marked off in degrees and are quite easy to lower to the exact temperature you want. Others, however, have controls marked "high" and "low," with a variety of settings in between. If yours is one of these, drop the control back a couple notches and then test the water temperature with a meat thermometer. If it is still too high, lower it another notch, and so on until you get the temperature you want. This is a no-investment measure—it costs you nothing but a few minutes of your time and begins saving you money at once.

◆ **Insulate your water heater.** Water heaters contain insulation, but if you can feel heat by putting your hand on the unit's side, then that heat—which should be making the water hot—is escaping from the unit. A unit that is warm to the touch can benefit from additional exterior insulation.

This is one of the easiest money-saving ideas; it is a quick, inexpensive, do-it-yourself project that will cost you about $20 to $25 for an insulation kit and half an hour of your time. It can save you its purchase price in the first year of operation. The kits, available in hardware and building-supply stores, contain a fiberglass overcoat for the water heater. The "coat" can be used on any kind of water heater, but follow the manufacturer's instructions to the letter to eliminate the risk of fire. The vent or flue at the top of gas and oil-fired heaters should not be blocked, nor should the control and terminal boxes on electrical heaters or you risk overheating the wiring.

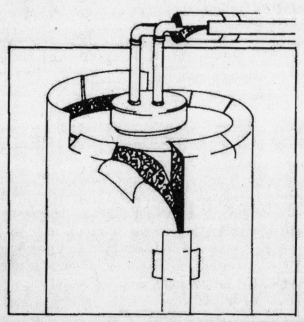

*Insulate your hot-water heater and hot-water pipes.*

- **Insulate your hot-water pipes.** As the hot water flows through unheated areas of your basement to sinks and bathtubs, some heat is lost by convection and radiant action through the pipes. You can prevent this loss by insulating your hot-water pipes. (You should also insulate cold-water pipes to eliminate condensation.) Home-improvement centers and hardware stores carry insulation designed specifically for water pipes. In most cases, you'll have to wrap fiberglass or a similar material around the pipes and then cover it with plastic or some other airtight material. There are also kits of pre-formed foam insulation that snap around the pipes. This type is far quicker to install, although it is more expensive.

- **Drain sediment from your water heater.** Sediment at the bottom of the tank in your hot-water heater acts as an insulation barrier between your heat source and the water—cutting your unit's efficiency. Every couple of months, drain off a gallon or two of water from the valve at the bottom of the tank to eliminate this sediment.

- **Clean heating elements once a year.** That same sediment can corrode the heating elements in your electric hot-water heater, making them less efficient. Once a year, cut the power to the unit, close the inlet valve, drain the tank, and carefully remove the heating elements (there should be two units, one at the middle of the tank, the other at the bottom). Clean off the

*Drain sediment from your hot-water heater.*

corrosion with a wire brush and replace the elements. To keep the elements from overheating, be sure you let the tank fill with water before you turn the electricity back on.

> ### ▼ Caution
>
> When insulating your hot water heater, remember (in Chapter 2, "Insulating Your Home") the insulation handling precautions: Wear long sleeves, a dust mask, and gloves.

- **Install an automatic flue damper.** In gas- and oil-fired water heaters, a lot of the flame's heat can be lost up the flue, just as it can be in a furnace. The remedy is the same—have an automatic flue damper installed. They are less expensive than those for a furnace—running about $150 to $200—and they should save you 10 to 15 percent of your water-heating costs. This is a job for an expert contractor; do not attempt to do it yourself.

- **Fix those leaks.** In addition to making a maddening drip-drip-drip noise all night, a leaky hot-water faucet can cost you money. If you've got a drop-per-second leak, then you're losing 3,120 gallons of hot water a year as well as the money it took to heat those gallons. Replacing a washer (the most common cause of a faucet drip) costs practically nothing and takes very little time.

- **Add aerators and flow restrictors.** Aerators introduce air into the flow of water at the faucet. While you perceive an undiminished flow of water and substantial pressure, actually less water is being used, so you save on heating fuel for the water you are *not* using. Install aerators and flow restrictors in all sinks and showerheads. (Some drought-prone jurisdictions prohibit the sale of any faucets or showerheads except those with aerators and flow restrictors.)

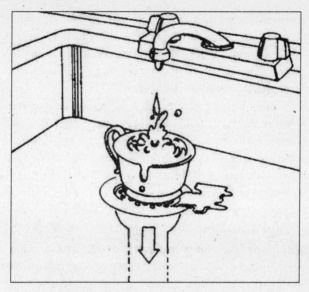

*A drip that fills a cup every 10 minutes wastes more than 3,000 gallons of hot water a year, plus the energy it took to heat that water.*

- ◆ **Take showers.** As long as we're in the shower, let's take one. Showers use less water than baths (provided you don't take inordinately long showers). Encourage your family to take showers to save on hot water. (And if you really want to save, try taking a "Navy shower." Aboard ship, where fresh water is at a premium, sailors wet themselves down and turn off the shower while they soap up. Then they turn the shower back on to rinse. In a Navy or even a civilian shower, the faster you move, the more you save.)
- ◆ **Fill those washers.** It takes as much hot water to run a dishwasher or clothes washer through a cycle when it is half loaded as it does when it is fully loaded, so wait until you've got those appliances filled up before running them. You'll do fewer washes, save on hot water, and save on electricity, too. (Also, your machines will last longer because they'll run less.)

- **Cold-water washes.** Wash clothes in cold water whenever possible. Or use a warm wash, but avoid a hot-water wash. For a dramatic comparison, look at the chart at the beginning of this chapter and compare the kWh usage of a washing machine using hot water—2,500 kWh—with the washing machine alone—that is, using cold water—103 kWh! Use the proper detergent and you'll still get clean clothes. A lot of newer garments can be rinsed in cold water, so save by shutting off the hot water in the rinse cycle when you can. And when you're buying clothing, read the care label and seek out cold-water washables. (No-iron garments save you the electricity you'd use on ironing—not to mention the dismal drudgery of the chore.)

- **Buy the right water heater.** When your old water heater wears out, buy the right size of replacement for your home. Too much capacity means too much wasted energy. If you've got a three-bedroom, two-bathroom home and you buy an 80-gallon unit, then you're wasting money, both when you purchase it and when you operate it. A 30-gallon unit is usually adequate for a two-bathroom house.

  If you aren't getting enough hot water from your present unit, you may not have to replace it. Frequently, you can save by buying a same-size unit and hooking it up parallel to your existing heater. (For example, two 30-gallon heaters may be more efficient—and less expensive—than a single new 60-gallon unit.) The best part about having two units in parallel is that if one goes down, you have a back-up until you can buy a replacement.

- **Install a pre-storage tank.** Your water heater is raising the temperature of water right from pipes under the street or from your well, which means it's beginning to heat water as cold as 40 degrees. Why not install an uninsulated storage tank downstream of the heater, which will enable the outside water to warm up to room temperature, approximately 70 degrees (even in a basement) before it goes into the water heater? This technology is called "ambient pickup."

## Chapter 5: Stretching Your Energy Dollar 143

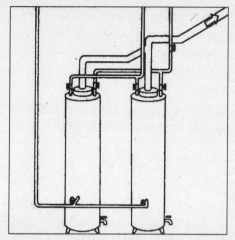

*Two hot-water heaters in series may save you money.*

 **Hint**

**What About the Sun?**

The sun's radiant energy is free. The equipment necessary to convert it to water heating isn't. The present state of the art is such that a solar hot-water system can provide two thirds of your family's hot water. They cost upward of $3,000 to $4,000 installed.

Now that we've shed some light on heat, let's shed some light on.

### Let There Be Light—but Only If You Need It

Up to a quarter of your home electric bill goes to light up your house. Now, to save money, you could go around turning off all the lights, using candles, and fumbling around in the dark. But don't do it: It's dangerous.

It's a good bet that if you're like most homeowners, there are substantial savings you can make in lighting without living in the dark. A few simple rules will cost you nothing and save you something:

- Turn off lights when you leave a room. A 50-watt bulb left burning for a full year can cost you $40.
- Reduce overall lighting in nonworking spaces of your home. Use only two incandescent bulbs in three-bulb fixtures.
- Buy table lamps that use three-way bulbs of 50-100-150 watts, and use the lowest-level illumination for the task at hand.
- Avoid long-life incandescent bulbs in most fixtures—they're less energy efficient.
- Keep lamps, bulbs, and light fixtures clean—they'll shed more light and you'll be less tempted to turn on additional lamps.

### Be Lumen-Wise

If you're like most people, you buy your light bulbs by the watt. But wattage is a measurement of how much electricity a bulb is burning, *not* how much light it gives off. Light values are measured in *lumens*, not watts. And it is possible that you'll be able to get more lumens (more light) with some lower wattage bulbs than with some higher-wattage bulbs. That means more light, even though you're burning less electricity.

The lumen value of a light bulb is printed on the carton. So it pays to compare different models of bulbs made by one manufacturer to get the most lumens per watt.

### Hint

#### How Much Will I Save?

Your saving will vary, depending on your electric rate and on how much you use your light. But, for every 10-watt reduction, you will probably realize a yearly savings of $6 to $9 (plus the purchase price of all those second bulbs).

Maybe someday we'll all buy our bulbs by lumens, not by watts ("Hey, I need an 1800 lumen bulb—what's the lowest wattage that'll give me that much light?"), but for now it's buyer beware.

## Watt Did You Say?

A high-wattage bulb *can* be more economical. A single 100-watt bulb may well give off more lumens than two 50-watt bulbs (and you also save the price of the second bulb. The following table indicates, in general terms, how you can replace two smaller bulbs with a single, larger bulb, and get equivalent lumens while burning fewer watts. (Again, this is merely a guide and you must check the lumens value on the carton.)

## Incandescent Comparison Chart

| Bulbs to Replace (Watts Is What You Pay) | Average Lumens (Lumens Is What You Get) | Per Bulb Average Life Hours | Average Lumens for Two Bulbs | Replace with One High Lumen Bulb and Save | | Average Watt Savings (Big Bucks over Time) |
|---|---|---|---|---|---|---|
| 2 25-watt | 212 | 1,500 | 424 | 1 40-watt | 440 | 10 |
| 2 40-watt | 440 | 1,000 | 880 | 1 60-watt | 870 | 20 |
| 2 50-watt | 490 | 1,000 | 980 | 1 75-watt | 1,200 | 25 |
| 2 60-watt | 870 | 1,000 | 1,740 | 1 100-watt | 1,750 | 20 |
| 2 75-watt | 1,200 | 750 | 2,400 | 1 100-watt | 1,750 | 50 |
| 2 100-watt | 1,400 | 750 | 2,800 | 1 150-watt | 2,160 | 50 |
| 2 150-watt | 2,160 | 750 | 4,320 | 1 200-watt | 3,300 | 50 |

## Fluorescent Comparison Chart

| Fluorescent | Size | Lumens | Average Life Hours | Replaces One Incandescent | Watt Saving |
|---|---|---|---|---|---|
| **Compact Type** | | | | | |
| 15-watt | 5³/₁₆" | 800 | 10,000 | 60-watt | 45 |
| 17-watt | 6⁷/₁₆" | 650 | 9,000 | 50-watt | 33 |

*continues*

## Fluorescent Comparison Chart (continued)

| Fluorescent | Size | Lumens | Average Life Hours | Replaces One Incandescent | Watt Saving |
|---|---|---|---|---|---|
| **Compact Type** | | | | | |
| 20-watt | 5$^{13}/_{16}$" | 1,100 | 10,000 | 75-watt | 55 |
| 24-watt | 6$^{11}/_{16}$" | 1,400 | 10,000 | 75/100-watt | 51/76 |
| 28-watt | 6$^{5}/_{16}$" | 1,650 | 10,000 | 100-watt | 72 |
| **Tube Type** | **Length** | | | | |
| 13-watt | 12" | 500 | 7,500 | 25-watt | 12 |
| 14-watt | 15" | 600 | 7,500 | 50-watt | 46 |
| 18-watt | 24" | 1,200 | 12,000 | 75-watt | 57 |
| 19-watt | 30" | 1,300 | 12,500 | 75/100-watt | 56/81 |
| 25-watt | 36" | 2,100 | 20,000 | 150-watt | 125 |
| 32-watt | 48" | 2,900 | 20,000 | 200-watt | 168 |

### What Is a Watt and What Is a Lumen ...

In brief, a watt is what you pay for, while a lumen is what you get.

A watt is a unit of measurement of power. One horsepower equals 745.7 watts. Your electricity consumption, measured in watts, is monitored by a meter, which your utility company reads monthly to determine how large (or small) a bill you get. That glass-enclosed meter in your basement or outside your house has a disk in it that spins, recording your use of watts. The faster it spins, the more money you're spending and the more electricity you're using—or wasting—and the bigger your utility bill will be. It's up to you to slooooow down that spinning disc by using fewer watts. Shut off the lights, change your lighting fixtures, and you can be a watt wizard.

A lumen is a unit of measurement of light. The more lumens you get per watt, the more efficient (and economical) your lighting systems becomes. Using fewer bulbs, or more watt/lumen-efficient bulbs, the more you'll save on your electric bill.

## Fluorescent Lighting

Now that you understand lumens, you'll understand why fluorescent light fixtures and the newer, screw-in fluorescent replacement bulbs can save you a *lot* of money. Quite simply, fluorescents deliver up to four times more lumens per watt than incandescent bulbs.

For example, you can get the same lighting values from a 25-watt fluorescent tube as from a 100-watt incandescent bulb. Your saving would be 75 watts—or as much as $65 a year. So, in a couple of years you can pay for the fluorescent fixture.

And now you don't even have to install a fluorescent fixture to get fluorescent benefits—many bulb-makers are turning out mini-, maxi-, and compact-type fluorescent bulbs, which screw into incandescent sockets. These newer fluorescents range from and 15-watt minis or compact types, producing about 800 lumens (comparable to a 60-watt incandescent bulb), all the way up to a 28-watt fluorescents producing 1,650 lumens (comparable to a 100-watt incandescent bulb).

Linear fluorescent bulbs—or tubes—which have been around for many years, were hardly ever considered a household lighting device because the color of light they emitted was thought to be harsh and unflattering, but that has changed, and the color of light is easier on the eye.

Consider how much energy you can save with these tubes of light. The most common is the four-footer, which burns 32 to 40 watts and delivers 2,900 lumens. It replaces a 200-watt incandescent light bulb and saves you 160 watts. The fluorescent tube lasts about 20,000 hours. Put another way, if you left it on for 24 hours a day, it would last for more than 800 days, or two years and nearly three months. Sure, fluorescent bulbs cost more than incandescent bulbs, but they last 5 to 26 times longer. No question, it's a bargain.

## ⚡ Caution

All fluorescent bulbs and tubes contain small amounts of mercury, a toxic substance. Mercury from landfills can eventually work its way into groundwater. Contact your sanitation department to find out how your community recommends properly disposing of spent fluorescent bulbs. You've done your share for the environment by using them and saving energy; now do a little more and dispose of them safely.

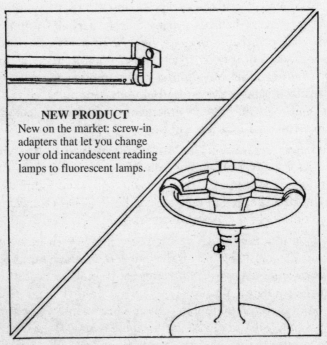

**NEW PRODUCT**
New on the market: screw-in adapters that let you change your old incandescent reading lamps to fluorescent lamps.

*Fluorescent ceiling fixtures will pay for themselves very quickly. The newer small fluorescent bulbs screw into lamps and other incandescent receptacles, and deliver more light per watt than conventional bulbs.*

## Here's a Switch

Actually, here are two switches—two switches that can save you money: a timer switch and a dimmer switch.

Timer switches cost as little as $15 and are extremely useful in basements, closets, or attics—areas of the house visited only occasionally, where a light could accidentally be left burning for many days—or even months.

These switches are easy to install. However, be sure you cut off the circuit supplying power to the switch before you begin working! Most common timers can be switched on for as long as an hour and will turn themselves off after that period. For children's playrooms or family rooms, you can buy four- or five-hour timer switches (the longer the maximum duration of time the switch allows, the more it will cost).

All dimmer switches are energy savers. Be sure you read the label of any dimmer you buy to see that it is Underwriters Laboratories listed.

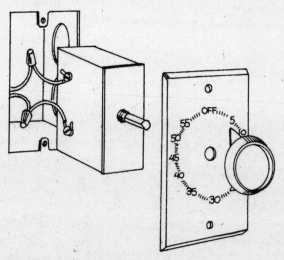

*Install a timer switch in seldom-used rooms and areas.*

# 150 Save Energy, Save Money

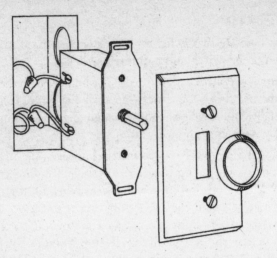

*Install energy-saving solid-state dimmer switches.*

## Daylight

It may sound obvious, but why not let the sun shine in? Sunlight's free, and if you open your curtains, drapes, shutters, blinds, and shades during daylight hours, you'll turn on fewer electric lights. (And keep those windows clean—a clean window, like a clean lighting fixture, gives you more light.)

## If You Can Stand the Heat, Stay in the Kitchen

Your kitchen appliances use a fair amount of electricity and/or gas. You can save money by following a few simple tips for each unit.

### Your Refrigerator

A refrigerator is a refrigerator, not a television set. There's no entertainment going on behind that door, so don't stand there with it open, gazing in and waiting for your favorite singing group to appear. And teach your children to open the door, take what they want quickly, and close it after them.

◆ **Dollar bill test.** If the gasket around the refrigerator door isn't tight, cold air escapes and the refrigerator's compressor motor has to work longer (which costs you more). Take a dollar bill, close it in the refrigerator, and see if it pulls out easily. If it does, your unit needs a new gasket or door adjustment.

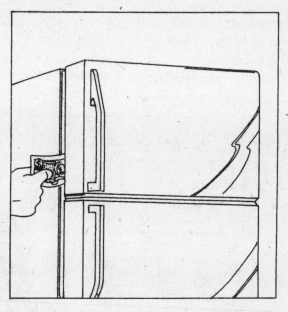

*If a dollar bill easily pulls out of your closed refrigerator door, your gasket is too loose. It must be replaced or the door adjusted.*

◆ **Defrost!** If you've got a manual defrost refrigerator, defrost it. A quarter-inch build-up of ice puts a sizable load on your compressor motor.

◆ **Buy two thermometers for your refrigerator.** Put one in the freezer, the other in the food compartment. Ideal temperatures are 40°F in the food area, 0 to 5°F in the freezer compartment. Any colder settings are simply a waste of money and crystallize the water in the food that causes freezer burn. Frozen fruit will have a taste change you will not like.

◆ **Buy the right refrigerator.** In the market for a new refrigerator? Check the EnergyGuide label and buy the kWh/Year model that uses the lowest possible amount of kilowatts, 800 or lower. Remember, models with Energy Star label exceed economy standards set by the government. Refrigerators with the freezer on top or on bottom are more economical to operate than those that have the freezer alongside the food compartment. Also, ice-making refrigerators use more electricity. And, when installing your new refrigerator, put it in the coolest spot in the kitchen to reduce its workload as much as possible. Avoid placing it near the oven.

*Vacuum the condenser coils at the back of the refrigerator at least once every six months. (Some refrigerator coils are at the bottom of the unit. These, too, should be vacuumed every six months.)*

## Your Range

When cooking, don't allow the gas flame to burn up around edges of a pot—you're wasting precious gas. The tips of the flame are the hottest part of the fire, so keep them *under* the pot. With electric ranges, use small coils for small pots.

- **Put a lid on it!** When you want to boil water, put the lid on the pot. It'll boil faster and cost less in cooking gas or electricity.
- **Keep your range top clean.** Clean reflectors and burners work more efficiently.
- **Cook sensibly.** When cooking on an electric range, turn off the power before the cooking time is up. The coils will retain enough heat to finish the job and you won't use as much electricity.
- *Steam,* don't *boil,* **vegetables.** It uses less energy and they taste better and retain more nutrients.
- **Use a microwave.** A microwave oven uses less energy than an electric oven. Besides, they cook faster.
- **Use your range selectively.** Small electric appliances such as toaster-ovens, electric fry pans, and grills are more economical for preparing small meals and snacks than an electric range or oven. Use them whenever possible.
- **Don't look.** Your oven isn't a television set. Don't keep popping the door open to look for entertainment—you'll just be wasting the heat and slowing down the cooking time and, in some recipes, ruining the meal.
- **Cook items together in the oven.** Prepare tomorrow's casserole with today's roast and you'll only have to cook once, instead of twice. (That'll save *your* personal energy, too.)
- **Buy a new range.** All new gas ranges by law have electronic ignition systems, but older ranges use about one third of their gas consumption to keep the pilot light burning—after all, they're on 24 hours a day, seven days a week. But if

you have an old range equipped with pilot lights, *do not turn off the pilot light* and light your stove with matches. This is extremely dangerous. Instead, consider buying a new range—you'll save energy.

 **Caution**

> Some dealers still have old, pilot-light models in their warehouses. You don't want one of these. Be sure you're buying a model with an electronic ignition system. If you're in the market for an electric range, check the EnergyGuide label and buy an economical one. And remember, the Energy Star means you're getting a range that's way above average in economy. Also, consider buying a unit with a built-in microwave oven.

### Washing Dishes

Here are some tips for saving money while dealing with your dirty dishes:

- **Don't run hot water.** When washing by hand, don't run hot water over all the dishes—it wastes energy. Fill the basin, wash the dishes, and then rinse.
- **Air-dry your dishes.** In your automatic dishwasher, use the air-dry cycle rather than the electrically heated drying cycle. If your dishwasher has no such cycle, open the door after the rinse cycle and let the dishes dry by evaporation.
- **Fill up your dishwasher.** Use your dishwasher only when it's fully loaded. If it's only half full, but there are no clean coffee cups or glasses left in the cupboard, take one out and wash it by hand. Don't be lazy; you can save a lot by running the washer only when it's full.

### Doing Laundry

Here are a few tips to save energy (and money) any day of the week in the laundry:

- **Fill that washer.** One full load uses half the energy of two partial loads—both in hot water and in electrical current to run the machine. Also, your washer will last longer if it runs fewer times per year.
- **Go for the cold.** Try to do cold- or warm-water washes, rather than hot-water washes. Some garments can be rinsed in cold water. If you have clothing like that, don't waste hot water on it.

### Caution

A number of devices are on the market that divert the hot air from your dryer's exhaust into your home. They also divert lint into your home and, in gas models, could present a carbon dioxide danger. These items are just plain *dangerous*. In addition to the carbon dioxide, the lint could be noxious, toxic, and even explosive. A dryer is for drying clothes, not for heating homes.

- **Pre-soak your laundry.** Pre-soak extremely dirty clothing, so you won't have to wash twice.
- **Take out the ol' clothes line.** Hang your wash out to dry on a clothesline on nice days. You'll save energy and your wash will smell fresher.

How much energy can these steps save? How much are you wasting now if you're not following them? If you really want to know, look into …

### Energy Monitors

The energy monitor is a device that can be installed in your kitchen or basement and can tell you moment-by-moment how much electricity your household is using in terms of cost per hour and total consumption. It's a little like pumping your own gas and watching the price go up as the tank gets filled. A monitor costs about $500 but it can probably *scare* you into cutting back on your consumption enough to pay for itself in a couple of years.

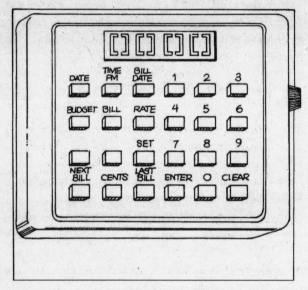

*An energy monitor.*

## Saving in Your Home Office

Increasingly, we find ourselves working in home offices. Whether it's our primary office or the spot where we catch up with work that we can't get to at a regular office, we are finding ourselves filling our homes with costly, electricity-consuming office machines like computers, faxes, scanners, printers, and copiers.

How can we save energy in our home offices?

◆ **Turn off the computer.** When it's not in use, turn it off; don't just put it to sleep. A sleeping computer is still drinking at the electrical trough.

- **Minimize the use of peripherals.** Minimize the use of your copier and scanner—they're real energy hogs. If you are copying two newspaper articles, try to get them into a single pass with your copier. Print multiple copies of documents rather than making copies with your copier—your printer uses less electricity than your copier.
- **Use e-mail rather than your fax machine.** A fax machine uses a good deal of electricity and—if you're sending a document across the country or to Lower Slobovia—you'll pay long-distance charges. E-mail's a local call.

### Hint

Because of the threats of brownouts and blackouts, you'll want to invest in a surge protector for your home office and any other electronic equipment you have. A surge protector stops surges of unusually high voltage from burning out your equipment. They cost from $30 up to about $150, but can save you thousands—and a lot of grief. One important consideration: If you're buying multiple surge protectors, buy them all from the same manufacturer. Mixing surge protectors often voids the manufacturer's protection guarantee.

- **Turn off the lights.** Light only those areas of your home office that need illumination, not the whole room. A single lamp with a fluorescent screw-in fixture on or beside your desk may be all the light you need to accomplish your home-office tasks.

And another thing: What about all those little glowing numbers on your gadgets—the fireflies of the electronic age? You know, the clock (usually flashing 12:00) on your VCR, the little green light on your CD player, the clocks on your dishwasher, microwave, electric oven, even on your TV set. Aren't they using juice?

Yes, they are. Perhaps as much as $100 a year worth of electricity powers those LEDs (Light-Emitting Diodes).

But before you go around unplugging all those gadgets to save the $100, consider this: Those LEDs represent electronic controls that have replaced mechanical controls (remember the old twist tuner for the TV set and the manual timer on the original microwaves?). These modern electronic controls are far more reliable than those old mechanical units, afford you much greater convenience, and, ultimately, save you money by making your appliances last longer. So learn to live with LEDs.

## Saving Behind the Wheel

For those who wax nostalgic about gasoline prices of $1.00 or $1.25, now that they're paying $30 or more to fill their tank, here's a sobering thought: When the baby boomer generation was sitting on daddy's lap, helping to steer the old family Chevy in the 1950s, gas cost about *20 cents* a gallon!!!

Well, we're about as likely to see that 20-cents-a-gallon gas in the future as we are to see $1.25 a gallon.

That said, unless you do an awful lot of driving, the gasoline portion of your annual energy expenditure is a lot less than your home heating and—in many cases—than your home appliance costs.

Gasoline prices get a lot of media attention for a very simple reason: It's a lot easier to send a television crew to a gas station and interview five or six disgruntled motorists than it is to videotape an inefficient gas furnace chewing up consumer dollars. The pictures drive the story and, in this case, distort the reality of the energy situation. But, we should all make every effort we can to save energy in every aspect of our lives, and we can help you save some of that precious gasoline money, so here goes:

# Chapter 5: Stretching Your Energy Dollar — 159

*No comment.*

The first, simplest, and most effective solution is this: don't get behind the wheel at all—*don't drive!*

- ◆ **Walk.** How many store and household tasks a week can be consolidated? How many destinations can you reach on foot? (The exercise will do your heart good, too!) Think about getting that old bicycle out of the garage, dusting it off, and using it for the short run to the post office or the stationery store. Many communities have created bike lanes to encourage folks to substitute pedal power for horsepower. Like walking, bicycling does your heart good.

- ◆ **Carpool.** More and more employers are offering incentives to workers who carpool or take public transportation. Take advantage of those incentives. If your employer doesn't offer them, put it in the suggestion box; mention that without the stress of driving every day, employees will arrive at work relaxed and they'll be more productive. (A bonus for riding buses and trains—you'll be able to read the newspaper while you're commuting, and learn about the latest gasoline price hikes that you don't have to pay).

Okay, you can't walk everywhere, and carpools and public transportation aren't always available. Sometimes you've just gotta drive. There are some things you can do to make that car of yours a little less thirsty. You can save about 10 to 12 percent or more of your gasoline costs if you do the following:

◆ **Drive slowly.** For the most part, the posted speed limits are the safest and the most fuel-economical speed limits. The faster you go, the more wind resistance you build up against your windshield, forcing your engine to work harder and increasing its gas appetite.

You'll also have more control over your car if you obey the speed limit; statistics indicate that over 65 miles an hour you're not steering your car, you're *pointing* it.

◆ **Avoid jackrabbit starts.** Sure, your car's faster than his car, but it'll cost you money to prove it. Just be content with the knowledge that you could whip him at the stoplight if you wanted to, but you're too smart for that kid stuff.

◆ **Avoid sudden stops.** If you race up to a stoplight and jam on the brakes, you're wasting gas. By coasting to that light you're taking advantage of the laws of physics: The car's forward momentum, rather than your engine, is propelling you to the light. That's our old friend, the law of inertia. If you let inertia help you, not only will you use less gas, but you'll also extend the life of your brakes and tires.

Economical driving techniques are also safe driving techniques. Follow these rules and you'll be less likely to get into an accident. You'll also do less harm to the environment because you'll be using less fuel.

# Chapter 5: Stretching Your Energy Dollar 161

*The stoplight Grand Prix is an expensive sport for the participants. Besides, it's downright dangerous.*

*Your engine, your wallet, your brakes, and your passengers will be grateful if you dispense with sudden stops.*

◆ **Keep your tires properly inflated.** Most cars list the correct tire pressure on a sticker on the side of the driver's door. Buy a tire-pressure gauge and check your pressure at least twice a month—more often after long trips, hitting large potholes, or scuffing your wheels against curbs. Low tires offer more resistance, so the engine must work harder to keep you up to speed. Over-inflated tires are *dangerous*—less rubber on the road means less braking and steering control. Having properly inflated tires also means they last longer—so you save money two ways.

◆ **Don't run your motor unnecessarily.** If you're going to be having a long conversation with your neighbor at the curb or you're going to run into the house for that umbrella you forgot, turn off the motor. Also, you don't need to "warm up" a modern engine for several minutes before you drive off. In times past, cars had to be warmed up to in order to work efficiently. Today, in most situations, half a minute is all the warm-up a car needs. (Make it a minute or two in extremely cold climates.)

*If you're going to be a motor mouth, give your car's motor a rest.*

Chapter 5: Stretching Your Energy Dollar **163**

◆ **Change your engine oil regularly,** at 5,000-mile intervals. Oil cuts friction, and friction makes your engine run inefficiently. Check your owner's manual to see if you need to change more frequently than 5,000-mile intervals.

◆ **Cruise if you can.** Use your cruise control on lightly traveled freeways and expressways; it'll keep you from creeping up to uneconomical speeds. (But, PLEASE, remain awake and attentive.)

◆ **Get it tuned.** Have your car tuned up at the mileage recommended in your owner's manual. A fresh set of spark plugs will give you a cleaner-running, more fuel-efficient engine. (If your gas mileage suddenly starts to drop off, or if you see excessive smoke from your tailpipe, it may be time for a tune-up.) A freshly tuned-up engine can save you as much as 5 percent of your gas costs.

*If your exhaust is belching smoke, it's trying to tell you something: "Tune me up!!!"*

- **Have your exhaust system checked annually.** Most modern cars use catalytic converters to control emissions. A defective converter—or an exhaust system with holes in it—can rob you of power and efficiency.
- **Get some fresh air.** Turn off your air conditioning and open the windows whenever possible. The air conditioner adds a considerable burden to your car's engine and costs you several miles per gallon. (In cold weather, the climate control does not impact fuel economy. Your car's heater doesn't put a strain on the engine, so in wintertime, don't freeze in the interests of fuel economy.)

### Fill 'Er Up

Which gas is "best"? The reality is, there's not much difference between brands—in fact, the different brands routinely swap product with each other. But are you buying more *octane* than you need? If your owner's manual recommends regular, you're not doing your car or your wallet any favors by putting in a higher-octane gasoline. Similarly, if your owner's manual recommends an octane rating between premium and the mid-level, try the mid-level first and see if your car "knocks" when you accelerate. (We don't know why they call it "knocking"; it sounds more like gurgling, and you can feel the vibration.) If your car knocks—or gurgles—go to the higher grade of gas. If it doesn't, you've just saved yourself about a dollar every time you fill up.

### Gas Price Watch

There is a Web site that lists the lowest-priced gasoline regionally around the country. It is www.gaspricewatch.com, but before you go driving 20 miles to save two cents a gallon, you need to figure out if the gas your jalopy burns getting you to the more economical station is going to cost more than you'll save on the fill-up. (You'll also want to calculate in the value of your time.) Remember, a dime a gallon savings is only $1.50 if you're filling up a 15-gallon tank; not even the cost of a single gallon at today's prices. However, www.gaspricewatch.com also tracks prices for home heating oil, and since it's the heating oil truck—not you—that does the driving around, you might fare better by comparison shopping for heating oil online.

## Buying Economy

All of these measures can save you about 10 to 12 percent on your gasoline bills. Your biggest step in automotive fuel economy is the one you take at the showroom when you select your next car.

Ask yourself, "How much car do I really need?"

Buy your next car with fuel economy in mind. Does it really take two tons of SUV to bring the kids to soccer practice? Or will a more economical minivan do just as well? Do you really need four doors, eight cylinders, and 250 horsepower to drive the three miles to work, or will a fuel-efficient two-door, four-cylinder economy car get you there in the same amount of time but considerably richer?

By federal law, every new car sold in the United States carries a price sticker with the Environmental Protection Agency's estimate of its city and highway gas mileage. Read that sticker, but bear in mind that, in our experience, the EPA's ratings have *always* been extremely optimistic; you're likely to get somewhat worse gas mileage in real life.

The EPA also offers a handy Web site that enables you to compare the relative fuel economy of specific cars, both new and used. That's at www.fueleconomy.gov.

The following chart indicates the annual cost (at a bargain basement $1.90/gallon) of driving a large luxury car or SUV, a standard sedan with an efficient four- or six-cylinder engine, and a super economy car (such as a gas/electric hybrid).

# 166 Save Energy, Save Money

The EPA's fuel economy estimates are a good indication of relative fuel economy, although they are usually overly optimistic.

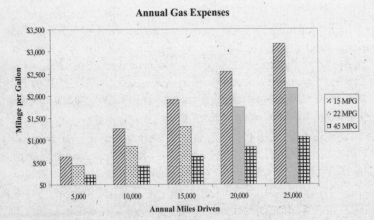

The annual cost of driving a large luxury car.

## Hybrids on the Horizon

One automotive solution to energy shortages and high fuel prices is hybrid cars. Currently there are two models on the market, and other manufacturers are racing to develop additional models. The U.S. Army is looking into hybrids for combat and support vehicles.

Simply put, hybrids overcome the disadvantages of battery-powered vehicles by taking advantage of the benefits of internal combustion engines. Powered by both gasoline and batteries, the current crop of hybrids—the Toyota Prius and the Honda Insight—charge their on-board batteries using the energy from the internal combustion engine, from forward motion and even from braking. Thus, the batteries never have to be plugged in to charge, and the cars get better mileage in city situations than they do on the open road. (Both city and road mileage figures are impressive, however: typically 50 mpg in the city and 45 over the road.) These vehicles are too new for us to know how long the batteries will last and, thus, what the actual cost per mile of the cars will be.

On the horizon are fuel cells to power vehicles. These use hydrogen and oxygen to create electricity. Current fuel cells are extremely large so they are best suited for trucks and buses. A fuel-cell-powered bus was tested at the Los Angeles International Airport. It was very quiet, used no gas or diesel fuel whatsoever, and its exhaust consisted of a slight drip of pure water. Current research is looking into creating the necessary hydrogen from ever renewable ethanol and methanol—which come from garbage, corn, biomass, or almost any organic source.

## Getting a Tax Credit

Uncle Sam, we want you ...

... to restore the energy tax credits.

# 168 Save Energy, Save Money

*"If you're not part of the solution, you are the problem."*

Two decades ago, the United States was in the grip of an international embargo by the oil-producing nations.

Gasoline prices soared to $1.00 a gallon and even at that outrageous price, there were long lines at the pumps and many frustrated drivers going home with only half a tank of gas. Home heating oil became too costly for some lower- and even middle-income families. The federal government lowered the thermostat in its buildings to 68 degrees in the winter and raised it to 78 degrees in the summer.

While the government did what it could, we wrote an energy-saving book telling individual homeowners what they could do to save energy, to save money, and—in those times—to save their country. At the time, the motivation to economize on our energy use was more than merely economic; it was patriotic.

It was, in the words of President Jimmy Carter, the "moral equivalent of war."

People responded. They bought fuel-efficient cars, cut the heat in the winter and the air conditioning in the summer, and demanded energy-efficient appliances.

The government gave Americans an economic incentive to conserve energy: It passed two categories of energy tax credits: Conservation Credits and Renewable Energy Source Credits.

These were not tax deductions, but credits—far more valuable. A tax credit means you take the amount you've spent and subtract it directly from the federal income taxes you owe.

The Conservation Credit was 15 percent of the first $2,000 spent on energy-saving items. That worked out to a maximum Conservation Credit of $300 and was applied to such improvements as insulation, caulking, weatherstripping, storm doors, and energy-efficient windows.

The second credit, the Renewable Energy Source Credit, was 30 percent of the first $2,000, plus 20 percent of the next $8,000 spent on such items as solar heating or electricity-generating windmills. The maximum Renewable Energy Source Credit worked out to a whopping $2,200.

What happened? There was an unprecedented amount of research and development into energy-efficient technologies. People did invest and did save energy. (Despite what you might hear, American homes are far more energy-efficient today than they were in the 1960s and 1970s.)

But, after the Carter administration, the energy tax credit was repealed. The oil-producing nations dropped their always-leaky embargo, and energy prices dropped. Public interest in energy conservation waned.

On a personal note, we doubted we would ever have to revisit the subject of energy conservation; an attitude of "burn it today, don't worry about tomorrow" seemed to dominate the public forums when it came to energy.

But, as Montaigne wrote, "Those who fail to study history are doomed to repeat it." A cavalier, if not disdainful, attitude toward energy conservation, has led us to a new national crossroad—and this time we've done it all by ourselves, without foreign blackmailers holding our economy hostage for political gain.

With this book, we've revisited the subject and offered tips on what you, as an individual, can do in your home to save energy and, hence, money.

We'd like to make a modest proposal to the government: Restore the Energy Tax Credit and substantially increase the amounts tax-payers can claim for both the Conservation Credit and Renewable Energy Source Credit.

Making a substantial credit available to taxpayers will result in major savings, which will have a huge positive ripple-effect throughout the economy.

Because they will be spending less on energy, consumers will have more to spend on other goods and services. By lessening dependence on foreign energy sources, the United States will be increasingly immune to blackmail and extortion. And by passing generous tax credits, the government will stimulate the development and manufacture of a whole range of unforeseeable technologies that will improve life for all of us for generations to come.

Our country is not just America the Beautiful, it's also America the Creative. Americans split the atom; Americans sent men to the moon; Americans landed robotic explorers on Mars, and Americans sent the first man-made object beyond our Solar System. We think—and know—that we have the technology to capture the sun's energy, exploit the tidal flows, harness the blowing winds so that we can be completely energy independent—without any environmental degradation.

## Photo-Voltaic Panels

One thing an energy tax credit might be used for would be to encourage homeowners (and the owners of apartment buildings, office buildings, and other commercial buildings) to install photo-voltaic panels.

What's a photo-voltaic panel? A panel that turns the free energy from the sun into electricity. Now don't be confused: This is not heating your domestic hot water with the sun's rays; this is generating electricity using the sun's rays. The technology, pioneered by NASA to power onboard computers and guidance systems on spacecraft has gotten quite sophisticated over the years.

For more than 30 years, the space agency has successfully used solar arrays to provide reliable energy to its far-flung spacecraft. Photo-voltaic panels have powered missions to Venus and Mars, and they power communications satellites in orbit around the earth. Some of these spacecraft have been in service for decades with no interruptions in power and no repairs to the system (think about the house call a solar-panel repair person would have to make to reach a satellite!!!). They are as reliable as the sunrise! Literally!

By themselves, photo-voltaic arrays can't supply all the power to all our homes and businesses, but they *can* make a difference; quite a significant difference.

Sometime in the future, when American know-how kicks in, the technology will enable every home to be energy independent. Can you imagine a world with no electric bills, no fuel bills? *Can you imagine* ...

Consider this: An array of 48 solar panels takes up only 503.53 square feet of roof space (about 29.5 feet by 17.33 feet). That solar array can—on average—generate 10,512 kilowatt hours of electricity a year at only 5 hours of sunlight per day. (Far more in sunny climates like the southwest.)

That output of power is as much electricity per year as we now get from 3.9 *tons* of coal or 15.6 barrels of crude oil.

Think of that. Your house can generate as much power as 3.9 tons of coal or nearly 16 42-gallon barrels (672 gallons) of crude oil. If your house was generating those kilowatts, those are tons of coal and barrels of oil that *would not* have to be taken from the ground. Additionally, by not burning those fossil fuels, our atmosphere would be cleaner and we'd be putting the earth at less risk of the disasters that are likely to accompany global warming.

Wouldn't you be willing to consider installing such a system if you got a tax credit for it?

And consider this: The U.S. Census Bureau found that there were approximately 88 *million* one-, two-, three-, and four-family structures in the United States. If every one of those homes was motivated to install photo-voltaic solar panels, they would generate 925,056,000,000 kWh of electricity (that is 925 billion). That would mean a savings of 343 million tons of coal that would not have to be mined and/or one billion, three hundred million barrels of crude oil that would not have be pumped out of the ground, refined, and burned.

A solar panel can probably work flawlessly for 35 to 50 years—perhaps even longer; they haven't been around long enough for us to know for sure. But NASA's case histories—in the hostile environment of deep space—are singularly encouraging. For homeowners, the only maintenance a photo-voltaic array would require might be the occasional hosing down to keep it clean.

Solar arrays aren't free. In fact, they're pretty pricey. But not only will they pay for themselves over time, but they'll help conserve finite resources (coal and oil), decrease air pollution, and—if enough of us install them—slow down the rapid pace of global warming.

All we need is some encouragement from the government in the form of energy tax credits. If Americans buy enough solar panels, the price will absolutely come down and will be affordable to all.

# Chapter 5: Stretching Your Energy Dollar 173

*Can you believe ... one billion, two hundred million barrels of crude oil that Americans won't have to buy?*

## Useful Energy Web Sites

All states and U.S. possessions have energy-conservation Web sites. They vary greatly in usefulness. Check out your state's site and if you feel it doesn't serve your needs, e-mail your governor and ask that it be upgraded. If enough people raise their e-voices, there will be improvement.

Of course, there's nothing to prevent a resident of one state from cyber-crossing state lines and accessing the energy Web site of a state with excellent information. We have marked with an asterisk (*) those state sites we consider worthwhile.

Remember, when visiting a site from a state other than your own, any energy tax incentive programs for residents of your state will be missing.

## State and Possession Web Sites

| | |
|---|---|
| Alabama Energy Office | www.naseo.org/members/states/alabama.htm |
| Alaska Energy Office | www.aidea.org/aea.htm |
| America Samoa Energy Office | www.naseo.org/members/states/asmsomoa.htm |
| Arizona Energy Office | www.azcommerce.com/energy.htm |
| Arkansas Energy Office Team | www.aedc.state.ar.us/energy/index.html |
| *California Energy Commission | www.energy.ca.gov/ |
| Colorado Energy Office (NASEO) | www.naseo.org/members/states/colorado.htm |
| Connecticut New Energy Technology (NET) | www.opm.state.ct.us/pdpd1/grants/net.htm |
| Delaware Div. of Public Advocate (link) | www.state.de.us/publicadvocate/dpa/html |
| District of Columbia (DC) Energy Office | www.dcenergy.org |
| Federated States of Micronesia Energy Office | www.visit fsm.org/index.html |
| Florida Div. of Housing and Community Development | www.dca.state.fl.us/fhcd/programs/sep/ |
| Georgia (energy saving, no document) | www.state.ga.us |

## Chapter 5: Stretching Your Energy Dollar 175

| | |
|---|---|
| Guam Energy Office | www.naseo.org/members/states/guam.htm |
| Hawaii Energy, Resources, and Technology Division | www.hawaii.gov/dbedt/ert/sitemap.html |
| Idaho | www.state.id.us |
| Illinois | www.state.il.us/state/sitemap.htm |
| *Indiana | www.state.in.us/doc/energy/factsfigs.html |
| Iowa Energy Office | www.naseo.org/members/states/iowa.htm |
| Kansas | www.accesskansas.org/ |
| Kentucky Division of Energy | www.nr.state.ky.us/nrepc/dnr/energy/doewebsites.html |
| Louisiana | www.state.la.us/ |
| *Maine | www.state.me.us/spo/energy/index.ht |
| Maryland Energy Administration | www.mdarchives.state.md.us/msa/mdmanual/25ind/html/33energh.html |
| Massachusetts Commonwealth Energy/Utilities | www.state.ma.us/ene.htm |
| Michigan | www.state.mi.us/ |
| Minnesota Department of Commerce | www.commerce.state.mn.us/ |
| Mississippi (no energy site found) | www.state.ms.us |

| | |
|---|---|
| *Missouri Department of Natural Resources | www.dnr.state.mo.us/de/efc.htm |
| Montana | www.state.mt.us/ |
| Nebraska | www.state.ne.us/ |
| Nevada Energy Office, Department of Business and Industry | www.energy.state.nv.us/ |
| New Hampshire, Governor's Office of Energy and Community Services | www.state.nh.us/governor/energycomm/tips.html |
| New Jersey | www.state.nj.us/ |
| New Mexico Energy Conservation and Management Div. | www.emnrd.state.nm.us.ecmd |
| *New York Energy Smart | www.getenergysmart.org |
| North Carolina | www.energycodes.org/states/nc/.htm |
| *North Dakota Energy Programs | www.state.nd.us/dcs/energy/default.html |
| North Mariana Islands | www.naseo.org/members/states/north_mariana.htm#top |
| *Ohio Office of Energy Efficiency | www.odod.state.oh.us/cdd/oee/default.htm |
| Oklahoma Energy Resources Board | www.oerb.com |
| *Oregon Department of Energy | www.facilities.das.state.or.us/Energy_conserv.html |

| | |
|---|---|
| Pennsylvania Energy Office | www.phmc.state.pa.us/DAM/rg/rg63.htm |
| Rhode Island | www.state.ri.us/dem/programs/index.htm |
| South Carolina | www.myscgov.com/SCSGPortal/static/home_tem1.html |
| South Dakota | www.state.sd.us |
| *Tennessee Energy Division | www.state.tn.us/ecd/energy.htm |
| Texas | www.sustainable.doe.gov/database/958.html |
| Vermont Environmental Board | www.state.vt.us/envboard |
| *Virginia Energy Handbook | www.mme.state.va.us/De/energybook/hbchap1.html |
| Washington Energy Facility Site Evaluation Council | www.efsec.wa.gov |
| West Virginia Department of Environmental Protection | www.dep.state.wv.us/ |
| Wisconsin | www.wisconsin.gov/state/home |
| Wyoming | www.state.wy.us |

## For More Information

For further information, check the following Internet resources:

State Energy Offices that provide resources for schools to save energy and teach about energy saving can be found at www.eren.doe.gov/energysmartschools/help_resources.html.

| | |
|---|---|
| Environmental Protection Agency Fuel Efficiency Web Site | www.fueleconomy.com |
| American Automobile Association | www.aaa.com |
| U.S. Department of Energy | www.doe.gov |
| Energy Smart Buildings and Other Sites | www.eren.doe.gov/buildings/rebuild/sitemap.htm |
| U.S. Energy Office | www.eren.doe.gov/buildings |

# Appendix A

# Track Your Energy Expenses

We have supplied you with the necessary information to save money—lots of money—on your energy bills. Now here's a way of keeping track of your progress. Use the tables in this appendix to chart your expenditures from month to month. As you implement more of our suggestions, you'll see an improvement from month to month. If you are keenly aware of how much you're spending, you'll be a more conscious—and conscientious—energy consumer in the future.

## Monthly Energy Expenses

### Before Energy-Saving

| 20__ | Jan. | Feb. | Mar. | Apr. | May. | Jun. | Jul. | Aug. | Sep. | Oct. | Nov. | Dec. |
|---|---|---|---|---|---|---|---|---|---|---|---|---|
| Electric | | | | | | | | | | | | |
| Gas | | | | | | | | | | | | |
| Fuel | | | | | | | | | | | | |
| Other | | | | | | | | | | | | |

## Monthly Energy Expenses

### After Energy-Saving—See What You Save

| 20_ _ | Jan. | Feb. | Mar. | Apr. | May. | Jun. | Jul. | Aug. | Sep. | Oct. | Nov. | Dec. |
|---|---|---|---|---|---|---|---|---|---|---|---|---|
| Electric | | | | | | | | | | | | |
| Gas | | | | | | | | | | | | |
| Fuel | | | | | | | | | | | | |
| Other | | | | | | | | | | | | |

# Appendix B

# Prepare for the Worst

This may read like a scenario out of a Hollywood action adventure film, but let's look at the unheroic, practical side: It's winter. A blizzard has struck, and you're stuck in your home. The power lines have gone down and there is no electricity. The electrical problem has resulted in a disruption of natural gas and—for your oil and propane customers—the blizzard has halted deliveries and you're out of fuel. It's 15 degrees outside and the thermometer inside is dropping fast.

Are you prepared for an all-encompassing emergency like that? In this appendix, we'll show you how to survive the worst possible conditions.

## Plan Ahead

Every home should have the following equipment:

- **An emergency food and water supply.** The food should be canned, jarred, or boxed goods that need no refrigeration. Buy high-energy foods (raisins, other dried fruit, honey, and so on) to help you keep warm, paper plates, and cups (every meal will become a picnic).
- **First-aid kit.** A first-aid kit consisting of compressed bandages of various sizes, plain gauze, a tourniquet, scissors, tweezers, and wood splints, and a first-aid manual.
- **Alternate lighting equipment.** Flashlights, extra batteries (keep batteries in plastic bags in your freezer to extend their shelf life), candles, matches, and hurricane lanterns with kerosene. (Boy Scout flashlights are particularly good since

they can be used as lamps or hooked to a belt to free your hands while they light your way.)

- **Emergency phone numbers.** You need a list of emergency contact information, in case phone service hasn't been disrupted or is restored before the crisis ends.
- **Battery-powered radio with plenty of extra batteries.** If the phone lines are out, the radio will be your only contact with the outside world.
- **Blankets.** You'll need extra blankets and/or sleeping bags (the down-filled ones are the warmest).
- **An alternate heating source.** A wood-burning stove with an ample supply of seasoned wood; a clean, well-functioning fireplace; kerosene, gasoline, or propane space heaters with lots of fuel.
- **Fire-fighting equipment.** Fire extinguishers, buckets of sand, a shovel, an ax. The alternate heating sources should present no danger if properly used, but you must be ready to deal with accidental fires.
- **Filled prescriptions.** Don't forget to stash extra supplies of any prescription medicine for family members taking medication. Remember also to check the expiration dates of your filled prescriptions regularly—medication doesn't last forever.
- **Warm clothing.** Wool is best, and dressing in layers is more effective than bulky single garments. Wool hats are critical (we lose half our body heat through uncovered heads). Longjohns are worth their weight in oil.
- **Ample supplies of newspaper, clear plastic wrap, tape.** You'll need these for wrapping your water pipes to prevent freezing.
- **Anti-freeze.** Also have a supply of automobile anti-freeze handy for toilet tanks, dishwashers, hard-to-reach drains.
- **Cards, games, books.** You'll need something to keep you and your family from contracting cabin fever.

Gathering these items and taking these steps in advance will improve your ability to weather the ultimate storm.

## When It Happens

Okay, you're prepared. Now it happens. There's snow up to your shoulders outside and no power, no telephone, no heat. What do you do?

- ◆ **Pull the plugs.** Unplug electrical appliances, computers, and your furnace's electrical starter. You don't want them damaged by a power surge when the electricity comes back on line.
- ◆ **Keep your freezer closed.** Use only food from the refrigerator. This will keep spoilage to a minimum. (After the emergency, refreeze only those items that still have ice crystals on them. Otherwise, keep defrosted freezer foods *refrigerated* [not refrozen] or prepare them. If the crisis lasts too long, throw out possibly spoiled food. There is no sense risking illness by eating foods that may have gone bad.)
- ◆ **Avoid intoxicants.** Don't drink alcoholic beverages or take tranquilizing drugs. They'll slow down your metabolism and you'll feel colder.
- ◆ **Wrap your pipes.** Wrap any uninsulated water pipes in newspaper and cover the newspaper with plastic to keep them from freezing. Open all water taps a little to keep water running. It will waste water, but so will a burst pipe. If it gets below freezing in the house, you may have to shut off your water from its source and drain all pipes. If you're doing that, be sure you empty out drain traps and the hoses behind your washing machine. Put automotive antifreeze in your toilet bowls and tanks to keep them from freezing. You can pour a little into your dishwasher and into hard-to-reach drains. (Be sure to run the dishwasher without dishes after the emergency to purge the anti-freeze.)

- **Cover your windows at night.** Hang blankets over your windows at night to keep cold air out and the heat in. Take them down during the day to get the sun's radiant heat.
- **Snuggle up.** Shut off rooms and centralize family life around the alternate heating source.
- **Bundle up.** Dress warmly and wear a wool hat—even when you sleep. Also, several light blankets are more effective than a single thick blanket. Bunch together to preserve body heat.
- **Stay indoors.** Try to stay dry, and try to stay indoors as much as possible.

## Emergency Lighting

You can make your candles last longer by putting them in a glass and filling the glass two-thirds full of cooking oil. When the candle burns down to the oil, it will turn into an oil lamp and should last for days as the oil burns.

You can make an emergency "Aladdin's lamp" from a creamer, a piece of cotton rag, and some peanut or other cooking oil.

Fill the creamer with oil, cut the rag into a wick with a piece hanging over the edge of the spout, and place the creamer on a larger dish, to prevent dripping. Light the wick and refill the oil as it burns down.

*Emergency lamps.*

## Exposure Treatment

Prolonged exposure to cold and wet weather can be serious, even life-threatening. Often, the person suffering exposure doesn't realize he or she is in trouble.

If you are exposed to too much cold and wet, get into dry clothing as quickly as possible. Apply a hot-water bottle or warm, dry towels to the trunk of your body. Raise your feet slightly to keep blood circulating to the brain. Drink warm drinks (but not alcoholic drinks). Remain quiet. Avoid massages or rubbing—they can do more harm than good.

Probably you will never face an emergency this dire, but even a lesser power outage or fuel shortage can be devastating if you're not prepared.

# Appendix C

# For More Information

For someone willing to do a little Web-surfing, there are virtually libraries full of energy-saving and money-saving information in cyberspace. We have supplied relevant Web sites at the end of each chapter. Here are all of them in one place, plus a few additional sites you may find extremely useful. The corporations, societies, and governmental agencies that supply invaluable information are marked with an asterisk (*).

| | |
|---|---|
| *Accurate Building Inspectors | www.AccurateBuilding.com |
| Air Conditioning and Refrigeration Institute | www.ari.org |
| American Architectural Manufacturers Association | www.aamanet.org |
| *American Society of Home Inspectors (ASHI) | www.ASHI.com |
| *A.O. Smith (Boiler & Heating Mfg.) | www.HotWater.com |
| American Society of Landscape Architects | www.asla.org |
| Andersen Windows (Makes Double-Glazed Windows) | www.andersencorp.com |
| *American Automobile Association | www.aaa.com |
| *California Energy Commission | www.energy.ca.gov |
| *Carrier Corporation (Air Conditioning Mfg.) | www.carrier.com |

| | |
|---|---|
| Casablanca Fan Company (Manufacturer of Ceiling Fans) | www.casablancafanco.com |
| Cellulose Insulation Manufacturers Association | www.cellulose.org |
| *Certainteed (Insulation Mfg.) | www.certainteed.com |
| *Consolidated Edison of New York (Electric Utility) | www.ConEd.com |
| *Davis & Warshow, Inc. (Plumbing Supply) | www.DavisWarshow.com |
| *Department of Energy, Office of Building Technology | www.eren.doe.gov/buildings |
| *Doody Home Centers | Hardware@DoodyHomeCenter.com |
| *Dykes Lumber Co. Inc. | www.DykesLumber.com |
| Energy Efficiency and Renewable Energy Clearinghouse | www.ornl.gov |
| Energy Smart Buildings and Other Sites | www.eren.doe.gov/buildings/rebuild/sitemap.htm |
| Energy Star | www.energystar.gov |
| Environmental Protection Agency Fuel Efficiency Web Site | www.fueleconomy.com |
| Fedders (Air Conditioners and Fans) | www.fedders.com |
| *Fisher/Merlis Television, Inc. | www.fishermerlistelevision.com |
| Gas Appliance Manufacturers Assn. | www.gamanet.org |
| *General Electric | www.GE.com |
| Heatilator Corp (Fireplace Inserts) | www.heatilator.com |
| *Home Depot | www.HomeDepot.com |
| *Honeywell, Inc. (Furnace, Air-Conditioning Thermostats) | www.honeywell.com |

# Appendix C: For More Information

| | |
|---|---|
| Hunter Fan Company (Ceiling Fan Manufacturer) | www.hunterfan.com |
| Insulation Contractors Association of America | www.insulate.org |
| *Keyspan (Brooklyn Union Gas) | www.KeyspanEnergy.com |
| National Association of Home Builders | www.nahb.com |
| National Fenestration Rating Council (Windows Standards) | www.nfrc.org |
| *NY State Energy Research and Development Authority | www.NYSERDA.org |
| National Wood Window and Door Association | www.nwwda.org |
| North American Insulation Manufacturers Association | www.naima.org |
| *Owens Corning (Insulation Manufacturer) | www.owenscorning.com |
| Pella Doors and Windows | www.pella.com |
| *Slant/Fin (Boiler Radiator Manufacturer) | www.SlantFin.com |
| *Trane (Air-Conditioner Manufacturer) | www.trane.com |
| *U.S. Census Bureau | www.census.gov |
| *U.S. Consumer Product Safety Commission | www.CPSC.gov |
| *U.S. Department of Transportation | www.DOT.gov |
| *U.S. Department of Agriculture | www.USDA.gov |
| *U.S. Department of Commerce | www.DOC.gov |
| *U.S. Department of Energy | www.doe.gov |

| | |
|---|---|
| *U.S. Department of Energy, Energy Efficiency Clearinghouse | www.eren.doe.gov/erec/factsheets |
| *U.S. Energy Office | www.eren.doe.gov/buildings |
| *U.S. Environmental Protection Agency | www.EPA.gov |
| *U.S. Department of Housing and Urban Development | www.HUD.gov |
| *U.S. Navy | www.NAVY.mil |

Business, especially big business, is often thought of as being environmentally indifferent, if not hostile; of being energy-mindless, if not wasteful. The private firms listed in this appendix fully recognize that conservation is not only good policy, it is also good business. They are fully aware that renewable, cheap—and, someday, free—energy can make them and their products highly competitive in the world market. They learned the essential lesson: One can do well by doing good.

# Index

## A

Accurate Building Inspectors, 187
aerators, faucets, energy efficiency, 140
AFUE (Annual Fuel Utilization Efficiency), 131
air conditioners
    central, SEER (Seasonal Energy Efficiency Rating), 131
    coils, 117-119
    economy runs, 119-120
    fans, 121-123
    filters, 117-119
    maintenance, 116-117
    motors, 117-119
    room, EER (Energy Efficiency Rating), 131
    shading, 120-121
air-conditioning
    cars, energy efficiency, 164
    energy quotient, 5-6
    ventilation, 116
air duct, return vents, 103-104
air systems, ventilation, 124
air valves, checking and replacing, 97
air vents, forced-air systems, 92
Alabama Energy Office, 174
Alaska Energy Office, 174
alcoholic beverages, power outages during blizzards, 183
America Samoa Energy Office, 174
American Society of Home Inspectors (ASHI), 187
American Society of Landscape Architects, 187
Andersen Windows, 187
anti-freeze, power outages, home equipment necessities, 182
appliances
    energy efficiency, 136-143
    kitchen, energy efficiency, 150-154
    laundry, energy efficiency, 154-155
    saving energy, 127-130
        reading labels, 131-136
    unplugging during power outages, 183
applying insulation, 18
Arizona Energy Office, 174
Arkansas Energy Office Team, 174
attics
    electrical conduits, 34
    installing insulation, 16, 30-34
        chimneys, 34
        pipes, 34
        trapdoors, 32
        walkways, 33
Automobile Club of America, 178, 187

## B

bands, energy savers, 58
bathroom, energy quotient, 7
baths, energy efficiency, 141
batteries (home equipment necessities during power outages), 181-182
batts, installing insulation, 18
    in attics, 30-32
blankets
    as insulation, 18
    home equipment necessity during power outages, 182
bleed hot-water systems, radiators, 98
blizzards
    exposure treatments, 185
    power outages
        emergency lighting, 184
        home equipment necessities, 181-183
        what to do, 183-184

boards, applying insulation, 19
boilers, AFUE (Annual Fuel Utilization Efficiency), 131
books (home equipment necessities during power outages), 182
British Thermal Unit. *See* BTU
broken windows, fixing, 46-47
  materials, 46-47
BTU (British Thermal Unit), 131
budgets, energy-saving, 9-11
bulbs, fluorescent, energy efficiency, 147-148

## C

California Energy Commission, 187
candles (home equipment necessities during power outages), 181
carpooling, energy efficiency, 159
cars
  hybrid, 167
  tune-ups, energy efficiency, 163
caulking, 54-58
ceilings, insulation, 17-18, 23
Cellulose Insulation Manufacturers Association, 188
central air conditioners, SEER (Seasonal Energy Efficiency Rating), 131
Certainteed, 42, 188
chemical insulation, 18-20
chimneys (attic), installing insulation, 34
clothes dryers, energy efficiency, 154-155
clothing, power outages
  blizzards, 184
  home equipment necessities, 182
cocklofts, insulation, 16
Colorado Energy Office (NASEO), 174
computers, energy efficiency, 156
Conservation Credits, governmental tax credits, 169
Consolidated Edison of New York, 188
construction areas, installing insulation, 41
copiers, energy efficiency, 157
crawlspaces
  floors above, insulating, 38-40
  insulating, 17-18, 24
cruise control, energy efficiency, 163
cutoff valve, steam radiators, 95-97

## D

Davis and Warshow, Inc., 188
daylight, energy efficiency, 150
Delaware Division of Public Advocate, 174
Department of Energy, Office of Building Technology, 188
derating
  gas furnaces, 100
  oil burners, 100
dimmer switches, energy efficiency, 149
dishwashers
  energy efficiency, 141, 154
  kWh/year (kilowatt-hours per year), 131
dollar bill test, refrigerators, 151
Doody Home Centers, 188
doors
  energy quotient, 5
  energy-savers, 43-44
  heating needs, 43-44
  insulating, 37
  weatherstripping, 70-71
drapes, energy-savers, 58
driving speed, energy efficiency, 160
dryers (clothes), energy efficiency, 154-155
Dykes Lumber Co. Inc., 188

## E

e-mail, energy efficiency, 157
E.Q. *See* energy quotients
economy, fuel, 165
economy runs, air conditioner, 119-120
EER (Energy Efficiency Rating), 131
efficiency (energy)
  appliances, 136-143
  daylight, 150
  energy monitors, 155-156
  government tax credits, 167-172
  home offices, 156-158
  kitchen appliances, 150-154
  laundry appliances, 154-155
  lights, 143-149
  photo-voltaic panels, 171-172
  solar panels, 172
  transportation
    gasoline, 158-165
    hybrid cars, 167

electrical conduits (attic), installing insulation, 34
emergencies
    blizzards, exposure treatment, 185
    phone numbers, 182
    power outages
        home equipment necessities, 181-183
        lighting, 184
        what to do, 183-184
energy
    appliances, reading labels, 131-136
    efficiency
        daylight, 150
        home offices, 156-158
        kitchen appliances, 150-154
        laundry appliances, 154-155
        lights, 143-149
        old appliances, 136-143
        transportation, 158-167
    government tax credits, 167-172
    monthly expenses, 179-180
    saving around the house, appliances, 127-130
    spending, 9-11
energy efficiency
    photo-voltaic panels, 171-172
    solar panels, 172
Energy Efficiency and Renewable Energy Clearinghouse, 188
Energy Efficiency Rating. *See* EER
energy monitors, 155-156
energy quotients (E.Q.), 1-2, 7-9
    air-conditioning, 5-6
    bathroom, 7
    doors, 5
    heating, 2-4
    insulation, 4-5
    kitchen, 7
    laundry room, 7
    lighting, 6
    television, 6-7
    ventilation, 5-6
    water heater, 3-4
    windows, 5
energy-savers
    bands, 58
    caulking, 54-58
    doors, 43-44
    drapes, 58
    landscaping, 76-79
    shades, 58
    storm doors, 61-62, 68
    storm windows, 61-62
    windows, 43-44
        tips, 44-45
Energy Smart Buildings, 178, 188
Energy Star, 42, 188
    label, insulation, 22
EnergyGuide labels, 132, 135-136
    ranges, 154
    refrigerators, 152
Environmental Protection Agency Fuel Efficiency, 178, 188
equipment
    home equipment necessities during power outages, 181-183
    installing insulation, 28-30
exhaust system check-up, 164
expenses, monthly, 179-180
exposure treatment, blizzards, 185
exterior walls, insulation, 17-18, 23
    installing, 35

# F

fans
    air-conditioner, 121-123
    temperatures, 121-123
faucets, energy efficiency
    aerators, 140
    fixing leaks, 140
    flow restrictors, 140
fax machines, energy efficiency, 157
Fedders, 188
filters
    changing, 93-94
    preventing blockages, cleaning, 93-94
fireplaces
    flue dampers, 104-106
    glass doors, 106-107
    power outages (home equipment necessities during power outages), 182
    wood-stove inserts, 107
first-aid kits (home equipment necessities during power outages), 181
Fisher/Merlis Television, Inc., 188
flashlights (home equipment necessities during power outages), 181
floors, installing insulation, 38-40

flow restrictors, faucets, energy efficiency, 140
flue dampers
   fireplaces, 104-106
   furnaces, 100-101
   water heaters, energy efficiency, 140
fluorescent bulbs, energy efficiency, 145-148
foam insulation
   applying, 19
   Urea Formaldehyde, 23
food (home equipment necessities during power outages), 181
forced-air systems, air vents, 92
freezers
   kWh/year (kilowatt-hours per year), 131
   power outages, blizzards, 183
fuel economy, 165
furnaces
   AFUE (Annual Fuel Utilization Efficiency), 131
   efficiency, 88
   flue dampers, 100-101

## G

games (home equipment necessities during power outages), 182
garages, insulating, 24, 40-41
Gas Appliance Manufacturers Association, 188
gas furnaces, derating, 100
gas-fired water heaters, Therms/Year, 131
gasoline, energy efficiency, 158-164
   fuel economy, 165
   gas price watch, 164-165
General Electric, 188
"ghosts," insulation, 17-18
glass doors, fireplaces, 106-107
government, energy tax credits, 167-170
   photo-voltaic panels, 171-172
Guam Energy Office, 175

## H

hanging baffles, old "pancake" boilers, 101-103
heat pumps, HSPF (Heating Seasonal Performance Factor), 131

Heatilator Corp., 188
heating, energy quotient, 2-4
heating ducts, checking and repairing, 92-93
heating elements, water heaters, energy efficiency, 139
heating needs, windows, 43-44
Heating Seasonal Performance Factor. *See* HSPF
heating systems, 81-83
   forced-air ducts, removing obstructions, 89-91
   motors, lubrication, 92
   radiators, 89-91
   removing obstructions, 89-91
   thermostats, 83-86
   tune-ups, 99
Home Depot, 188
home equipment necessities during power outages or blizzards, 181-183
home offices, energy efficiency, 156-158
homes, insulation, 13-14
   applying, 18
   attics and cocklofts, 16
   energy efficiency
      appliances, 136-143
      daylight, 150
      energy monitors, 155-156
      home offices, 156-158
      kitchen appliances, 150-154
      laundry appliances, 154-155
      lights, 143-149
   energy-saving appliances, 127-136
   exterior walls, 17-18
   installing, 27-41
   R-value, 14-16
   regulations, 21-22
   types, 18-20
   vapor barriers, 25-27
   where to insulate, 22-25
Honeywell, Inc., 188
HSPF (Heating Seasonal Performance Factor), 131
humidifiers
   hygrometers, 86-87
   thermostat, temperatures, 86-87
Hunter Fan Company, 189
hybrid cars, energy efficiency, 167
hygrometers, humidifiers, 86-87

## I–J

incandescent bulbs, energy efficiency, 145-146
inorganic insulation, 18-20
insulating material, weatherstripping, 52
insulation
    applying, 18
    attics, 30-34
    batts, 18
    blankets, 18
    boards, 19
    ceilings, 23
    chemical, 18-20
    crawlspaces, 24
    doors, 37
    energy quotient, 4-5
    ENERGY STAR labels, 22
    exterior walls, 23
    floors above crawlspaces, 38-40
    foam, 19
        Urea Formaldehyde (UF), 23
    garages, 40-41
    "ghosts," 17-18
    homes, 13-18
    inorganic, 18-20
    installing, 27-28
    kits, Underwriters Laboratories (UL), 36
    lighting fixtures, 32
    loose fill, 18
    motors, 32
    new construction areas, 41
    organic, 18-20
    regulations, 21-22
    tools and equipment needed, 28-30
    types, homes, 18-20
    unused rooms, 40-41
    vapor barriers, 25-27
    walls, 34-41
    water heaters, energy efficiency, 137-138
    water pipes, energy efficiency, 139
    where to insulate, 22-25
    window frames, 37
interlocking thresholds, weatherstripping, 76
Iowa Energy Office, 175

## K

Kentucky Division of Energy, 175
Keyspan, 189
kilowatt-hours, estimated household consumption, 127-130
kitchen
    appliances, energy efficiency, 150-154
    energy quotient, 7
kits, insulation, Underwriters Laboratories (UL), 36
kWh/year (kilowatt-hours per year), 131

## L

labels
    appliances, saving energy, 131-136
    EnergyGuide, 132-136
        ranges, 154
        refrigerators, 152
landscaping, energy savings, 76-79
lanterns (home equipment necessities during power outages), 181
laundry appliances, energy efficiency, 154-155
laundry room, energy quotient, 7
leaks, faucets, energy efficiency, 140
lighting
    energy quotient, 6
    power outages during blizzards, 184
lights
    energy efficiency, 143-144
        fluorescent bulbs, 147-148
        lumens, 144-146
        switches, 149
        watts, 145-146
    insulation, 32
living conditions, temperatures, 116
loose fill insulation, applying, 18
loose frames, fixing, 51
loose sashes, fixing, 49-50
lumens
    energy efficiency, 144-146
    fluorescent bulbs, 147-148

## M

maintenance, air conditioners, 116-117
Maryland Energy Administration, 175
masonry walls, insulating, 36-37
matches (home equipment necessities during power outages), 181
medications (home equipment necessities during power outages), 182
Minnesota Department of Commerce, 175
Missouri Department of Natural Resources, 176
monitors (energy), 155-156
monthly energy expenses, 179-180
motors, insulating, 32

## N

National Association of Home Builders, 189
  insulation, 22
National Fenestration Rating Council, 189
National Wood Window and Door Association, 189
New York Energy Smart, 176
newspapers (home equipment necessities during power outages), 182
North Dakota Energy Programs, 176

## O

octane, 164
offices (home), energy efficiency, 156-158
Ohio Office of Energy Efficiency, 176
oil burners, derating, 100
oil changes, energy efficiency, 163
organic insulation, 18-20
ovens, energy efficiency, 153-154
Owens Corning, 189

## P

panels
  photo-voltaic, 171-172
  solar, 172
Pella Doors and Windows, 189
phone numbers (home equipment necessities during power outages), 182
photo-voltaic panels, 171-172
  energy efficiency, 171-172

pipes (attic), installing insulation, 34
pipes (water), wrapping, 183
plastic sheeting, storm windows, 65-67
plastic wrap (home equipment necessities during power outages), 182
porches, insulation, 24
power outages during blizzards
  emergency lighting, 184
  home equipment necessities, 181-183
  what to do, 183-184
prescriptions (home equipment necessities during power outages), 182
pressure-sensitive foam, weatherstripping, 71-72
prestorage tanks, water heaters, energy efficiency, 142
printers, energy efficiency, 157

## Q-R

quotients, energy
  air-conditioning, 5-6
  bathroom, 7
  doors, 5
  heating, 2-4
  insulation, 4-5
  kitchen, 7
  laundry room, 7
  lighting, 6
  television, 6-7
  ventilation, 5-6
  water heater, 3-4
  windows, 5

R-value, insulation, 14-16
  installing, 27-28
radiators
  bleed hot-water systems, 98
  heating systems, 89-91
radios (home equipment necessities during power outages), 182
ranges
  energy efficiency, 153-154
  EnergyGuide label, 154
refrigerators
  dollar bill test, 151
  energy efficiency, 150-152
  EnergyGuide label, 152
  kWh/year (kilowatt-hours per year), 131

regulations, insulation, 21-22
Renewable Energy Source Credits, governmental tax credits, 169
resistance, R-value, insulation, 14-16
return vents, air ducts, 103-104
roofing materials, money savings, 76
room air conditioners, EER (Energy Efficiency Rating), 131
rooms, unused, insulating, 40-41

## S

scanners, energy efficiency, 157
Seasonal Energy Efficiency Rating. *See* SEER
sediment, water heaters, energy efficiency, 139
SEER (Seasonal Energy Efficiency Rating), 131
shades, energy savers, 58
showers, energy efficiency, 141
siding materials, money savings, 76
single-pane storm windows, 63-64
Slant/Fin, 189
solar panels, energy efficiency, 172
space heaters (home equipment necessities during power outages), 182
spending, energy, 9-11
spring-metal weatherstripping, 72
steam radiators
    cutoff valve, 95-97
    pitched properly, 95
storm doors
    energy savers, 61-62, 68
    selections, 68
storm windows
    combinations, 64
    energy savers, 61-62
    plastic sheeting, 65-67
    single-pane, 63-64
storms, power outages
    emergency lighting, 184
    home equipment necessities, 181-183
    what to do, 183-184
stoves, energy efficiency, 153-154
surge protectors, 157
sweeps, installing, 74-75
switches, lights, energy efficiency, 149

## T

tanks (prestorage), water heaters, energy efficiency, 142
tape (home equipment necessities during power outages), 182
tax credits, energy efficiency, 167-170
    photo-voltaic panels, 171-172
television, energy quotient, 6-7
temperatures
    fans, 121-123
    living conditions, 116
    thermostat, 83-86
        humidifier, 86-87
    water heaters, energy efficiency, 137
Tennessee Energy Division, 177
thermometers, refrigerators, energy efficiency, 151
thermostat
    heating systems, 83-86
    temperatures, 83-86
Therms/Year, 131
timer switches, energy efficiency, 149
tire pressure, energy efficiency, 162
tools, installing insulation, 28-30
Trane, 189
transportation
    gasoline, energy efficiency, 158-165
    hybrid cars, energy efficiency, 167
trapdoors (attic), installing insulation, 32
treatments, exposure, blizzards, 185
tune-up, heating system, 99

## U

U.S. Census Bureau, 189
U.S. Department of Agriculture, 189
U.S. Department of Commerce, 189
U.S. Department of Energy, 42, 178, 189-190
U.S. Department of Energy and Environmental Protection Agency, ENERGY STAR labels on insulation, 22
U.S. Department of Housing and Urban Development, 190
U.S. Department of Transportation, 189
U.S. Energy Office, 178, 190

U.S. Environmental Protection Agency, 190
U.S. Navy, 190
Underwriters Laboratories (UL), insulation, 22
   kits, 36
unfinished walls, insulation, installing, 35-36
United States Consumer Product Safety Commission, insulation, 22
unused rooms, installing insulation, 40-41
Urea Formaldehyde (UF) foam insulation, 23

## V

vapor barriers, insulation, 25-27
   installing in attics, 30-32
ventilation
   air conditioning, 116
   air systems, 124
   alternatives, 123
   energy quotient, 5-6
Vermont Environmental Board, 177
Virginia Energy Handbook, 177

## W–X–Y–Z

walkways (attic), installing insulation, 33
walls, installing insulation, 34
   exterior, 17-18, 23, 35
   masonry, 36-37
   unfinished, 35-36
washing machines
   energy efficiency, 141-142, 154-155
   kWh/year (kilowatt-hours per year), 131
water (home equipment necessities during power outages), 181
water heaters
   energy efficiency
      buying the right one, 142
      cleaning heating elements, 139
      draining sediment, 139
      installing an automatic flue damper, 140
      installing prestorage tanks, 142
      insulation, 137-138
      lowering water temperature, 137
   energy quotient, 3-4
   gas-fired, Therms/Year, 131
water pipes
   energy efficiency, insulation, 139
   wrapping, 183
watts
   fluorescent bulbs, energy efficiency, 147-148
   lights, energy efficiency, 145-146
weatherstripping
   aluminum-backed vinyl, 74
   doors, 70-71
   insulating material, 52
   interlocking thresholds, 76
   pressure-sensitive foam, 71-72
   spring-metal, 72
   wood-backed foam rubber, 73
West Virginia Department of Environmental Protection, 177
windows
   blizzards, 184
   broken, fixing, 46-47
   double-glazed, 68-69
      cost, 68-69
   energy quotient, 5
   energy-savers, 43-44
   energy-saving tips, 44-45
   frames, installing insulation, 37
   heating needs, 43-44
   latches, fixing, 48-49
      materials, 48-49
      tools required, 48-49
   saving money, 43-44
winter, heating, energy quotient, 2-4
wood-burning stoves
   economical, 114-115
   home equipment necessities during power outages, 182
   inserts, 107-108, 111
   safety tips, 111-113